Holding Back Disaster

Holding Back Disaster: Protecting Lives and Infrastructure Through Excavation Damage Prevention presents a rare system-level account of excavation damage prevention as a form of critical infrastructure, bringing together the people, systems, and practices that protect people and essential infrastructure every day.

Excavation damage to buried infrastructure is one of the most persistent, costly, and under-examined safety problems in the built environment. Each year in the United States alone, millions of excavation activities place buried gas, electric, water, sewer, and telecommunications assets at risk, resulting in fatalities, injuries, service outages, environmental harm, and billions of dollars in economic losses. The book traces the evolution of undergrounding and coordination practices from early utility cooperation through the emergence of One-Call/811 systems, modern locating technologies, and data-driven risk management. Rather than treating utility strikes as a series of isolated failures, the book demonstrates how outcomes are shaped by institutional design, incentives, workforce dynamics, data quality, and regulatory fragmentation. The book presents narratives and case studies, supported by real-world incidents, quantitative estimates, and expert perspectives.

This book is essential reading for infrastructure and construction practitioners and leaders who work across excavation, utilities, locating, safety management, and risk.

Scott Landes is an entrepreneur, inventor, publisher, and global conference founder with more than 40 years of leadership in underground infrastructure safety and damage prevention. He is the holder of 10 patents related to damage-prevention products and systems and is the former co-chair of the Common Ground Alliance Education Committee and recipient of the CGA Jim Barron Award.

Benjamin R. Dierker, JD, MPA, is the Executive Director of the Alliance for Innovation and Infrastructure (Aii), a nonprofit think tank focused on infrastructure systems, safety, resilience, and public policy. He is the author of more than 75 policy reports and a widely cited expert on infrastructure governance and risk.

"You will never look at spray paint on the sidewalk the same way again. This book illuminates the most vital profession you never knew existed and the life-or-death stakes of their work."

– **Jemmie Wang,** *Damage Prevention Expert*

"This book shines a needed light on a part of our infrastructure system that is essential but too often overlooked. By clearly explaining the risks associated with excavation and the systems in place to mitigate them, Benjamin R. Dierker and Scott Landes contribute meaningfully to a broader understanding of how safety, coordination, and accountability intersect in infrastructure delivery. *Holding Back Disaster* is a valuable resource for practitioners and policymakers alike, and a reminder that even as we invest in building more, we must remain disciplined in protecting what already exists."

– **DJ Gribbin,** *Founder, Madrus, LLC; Former Special Assistant to the President for Infrastructure Policy*

"As an author of many industry publications myself, I understand how difficult it is to tell the story in a convincing way. I am deeply impressed in the way this book weaves the daily safety impacts with the processes in a manner that is comprehensive and readable."

– **Louis Panzer,** *Executive Director NC811*

"*Holding Back Disaster* peels back the layers of safety exposing the hidden world beneath our feet and the extraordinary people working tirelessly and collaboratively to protect it and us. This book is a powerful reminder that safeguarding our infrastructure and our lives can pivot on a single act – one call or one click."

– **Mike Sullivan,** *President Utility Safety Partners*

"This book accurately captures the "heart of utility damage prevention." and provides a complete context of need and solutions that have saved so many lives and as a bonus has protected buried facilities."

– **Wayne Jensen,** *Vice President, Director of Safety at Stahl & Associates Insurance*

"The most interesting history of the Damage Prevention Industry, that is encapsulated in one place. If you want to make a difference and shape the future, this is the industry for you."

– **Duane Rodgers,** *CEO PelicanCorp*

"For those of you who have asked for the history of 811, this is the resource."

– **Roger Cox,** *President ACTS Now*

"Reads more like a mystery novel than a textbook. That's a compliment! Teaching by stories can be very effective."

– **Jim Hayes,** *President Fiber Optic Association (FOA)*

"Damage Prevention is one of those things that you may not realise that you need, until your life changes forever. The better we get at preventing damage, the more invisible the need becomes. But the importance of Damage Prevention is greater than ever – and growing – and that's why this insightful and entertaining book is needed now more than ever. Scott and Benjamin have drawn on their vast combined experience and network of contacts to create a fascinating account of the breadth, complexity and challenges of damage prevention not just in the US but across the world, but perhaps most importantly a call to action and some ideas of the new technologies and approaches that might lead us to the next big breakthrough."

– **Chris Ross,** *Principal – Asset Protection & Relocation at Telstra*

"Our Nation, along with other advanced countries, has a high standard of living due mostly to our utility and transportation networks. The importance of preventing damage to existing utilities, and finding room for new ones, cannot be overstated to maintaining this standard of living."

– **Jim Anspach, P.G. (r), Dist.M.ASCE, NAC, F.UESI,**
Affiliate Professor of Practice Iowa State University

"In a time when underground infrastructure is more critical than ever, this book shows how it is protected and how the damage prevention industry was built."

– **Nick Bonstell,** *President & CEO MISDIG 811*

"*Holding Back Disaster* translates the often invisible world of underground utilities into a clear, actionable framework for smarter infrastructure planning and system integration. Grounded in real cases, evolving technologies, and policy tools, it belongs on the syllabus of graduate programs focused on resilient infrastructure and the built environment."

– **Dr. Paul Dickinson, PhD,** *Founder, Principal Smart Infrastructure Solutions*

"I very much enjoyed the read and believe it would help those new to the Damage Prevention enterprise to get their bearings. Just remember, these accidents involved real – sometimes fatal – consequences to real people and property. That is a risk we can all live without!"

– **Jeff Wiese,** *Former Associate Administrator for Pipeline Safety*

"This is a fantastic overview of damage prevention in relation to underground utilities, with even a nod to overhead utilities. The breadth and scope is admirable, covering important and relevant themes and topics tied to damage prevention, and the book can offer key insights to both those in the industry and out."

– **Scott Crawford,** *President & CEO Virginia811*

"To know the industry's history is to understand not only its relevance but also the future decisions needed to maintain relevance. There isn't much difference between damage *prevention* and damage *promotion,* and this book goes a long way in proving just that. Congratulations to the authors for clearly stating that we will choose one or the other."

– **Mike Parilac, Founder,** *Staking University, Planet Underground TV*

"*Holding Back Disaster* brings overdue attention to the hidden infrastructure systems and damage prevention practices that protect lives, communities, and critical services every day. By blending real-world industry insight with a clear public-safety lens, this book offers a valuable and timely resource for engineers, policymakers, utility professionals, business leaders, and anyone concerned with infrastructure resilience."

– **Dr. Christopher B. Hill, PE, PhD, CRMP**, *Professor – Southern Methodist University, President & CEO – KC Hill Consulting*

"*Holding Back Disaster* reframes infrastructure safety as a problem of governance rather than engineering alone. Its emphasis on incentives, information problems, and coordination failures shows how legal and policy design shape outcomes in complex systems. A valuable resource for students and practitioners in law, economics, and public administration."

– **Andrew P. Morriss,** *Professor of Law and International Affairs, Texas A&M University*

"From an emergency management and response perspective, the ideal condition is preventing incidents from occurring. This can be best accomplished by emphasizing risk assessment, maintaining a strong safety culture, coordination of key stakeholders, and comprehensive maintenance of underground infrastructure. However, emergency incidents can occur so having well developed response plans and systems will minimize the consequences of these occurrences. *Holding Back Disaster* reinforces the importance of prevention and its impact on life safety, protection of property, and critical infrastructure."

– **Timothy P. Butters**, *Former Acting Administrator, PHMSA; Former Assistant Fire Chief and Emergency Management Leader*

"Reliable energy infrastructure depends not only on what we build, but how well we protect and maintain it. By highlighting the risks to critical underground systems and the importance of coordination across stakeholders, *Holding Back Disaster* brings timely attention to an often overlooked factor in infrastructure resilience and economic stability."

– **Neil Chatterjee**, *Former Chairman, Federal Energy Regulatory Commission (FERC)*

Holding Back Disaster

Protecting Lives and Infrastructure Through Excavation Damage Prevention

Scott Landes and Benjamin R. Dierker

NEW YORK AND LONDON

Designed cover image: Shutterstock

First published 2027
by Routledge
605 Third Avenue, New York, NY 10158

and by Routledge
4 Park Square, Milton Park, Abingdon, Oxon, OX14 4RN

Routledge is an imprint of the Taylor & Francis Group, an informa business

© 2027 Scott Landes and Benjamin R. Dierker

The right of Scott Landes and Benjamin R. Dierker to be identified as author[/s] of this work has been asserted in accordance with sections 77 and 78 of the Copyright, Designs and Patents Act 1988.

All rights reserved. No part of this book may be reprinted or reproduced or utilised in any form or by any electronic, mechanical, or other means, now known or hereafter invented, including photocopying and recording, or in any information storage or retrieval system, without permission in writing from the publishers.

For Product Safety Concerns and Information please contact our EU representative GPSR@taylorandfrancis.com. Taylor & Francis Verlag GmbH, Kaufingerstraße 24, 80331 München, Germany.

Trademark notice: Product or corporate names may be trademarks or registered trademarks, and are used only for identification and explanation without intent to infringe.

ISBN: 978-1-041-35208-2 (hbk)
ISBN: 978-1-041-35121-4 (pbk)
ISBN: 978-1-003-79661-9 (ebk)

DOI: 10.1201/9781003796619

Typeset in Times New Roman
by KnowledgeWorks Global Ltd.

This book is based on firsthand experience, expert interviews, and extensive research. The contents were peer reviewed by industry experts. Meet them in the acknowledgments.

This book is dedicated to:

The most vital profession you never knew existed.

Contents

Foreword

Infrastructure works best when it serves people – not systems, not bureaucracies, and not assumptions carried forward simply because they are familiar. At its core, transportation is about enabling safe, efficient movement and supporting the economic and social connections that define thriving communities. Yet too often, the systems we rely on to build, maintain, and protect that infrastructure fall short of that standard.

Excavation damage prevention is one of those areas where the gap between intention and outcome remains too wide. Every year, preventable incidents disrupt service, threaten public safety, and impose unnecessary costs on businesses and communities alike. These are not abstract risks – they are tangible consequences of misaligned incentives, inconsistent practices, and fragmented accountability across stakeholders who all share responsibility for the same physical space.

The challenge is not a lack of awareness. The tools exist. The processes exist. In many cases, the laws and regulations exist. What remains uneven is execution – and more importantly, the alignment of systems that ensure consistent, reliable outcomes.

That is why this book matters.

It approaches excavation damage prevention not as a narrow technical issue, but as a broader systems challenge – one that sits at the intersection of infrastructure governance, operational coordination, and human behavior. It recognizes that safety is not achieved through any single intervention, but through the cumulative effect of policies, practices, and decisions that either reinforce or undermine one another.

The 811 system represents one of the most practical and widely adopted tools in this space. It is a simple concept with significant impact – connecting excavators with facility operators to ensure that underground infrastructure is identified and protected before work begins. When used effectively, it reduces uncertainty, improves communication, and helps prevent costly and dangerous incidents.

But like any system, its effectiveness depends on how it is implemented and integrated into broader workflows. Variability across jurisdictions, differences in enforcement, and gaps in compliance can limit its potential. The result is a system that works well in some contexts and less reliably in others – not because the concept is flawed, but because the surrounding ecosystem is inconsistent.

This is where thoughtful analysis and practical guidance become essential.

One of the strengths of this work is its focus on understanding the underlying drivers of performance. It does not treat excavation damage as an isolated problem, but as the outcome of a complex network of decisions – from how projects are planned and scheduled, to how information is shared, to how accountability is defined and enforced. By examining these dynamics, the book provides a clearer picture of where improvements can have the greatest impact.

Equally important, it highlights the role of innovation – not as a buzzword, but as a necessary component of progress. Advances in data, mapping, and communication technologies offer new opportunities to improve accuracy, efficiency, and transparency. At the same time, innovation must be paired with governance structures that support adoption and ensure that new tools are used effectively.

Experience has shown that neither the public nor the private sector can address these challenges alone. Success depends on collaboration – on bringing together regulators, operators, contractors, and technology providers to align incentives and share responsibility for outcomes. When those partnerships are structured well, they can drive meaningful improvements in safety and performance. When they are not, fragmentation persists.

This book contributes to that collaborative effort by providing a framework for thinking about excavation damage prevention in a more integrated way. It encourages stakeholders to move beyond compliance as a minimum standard and toward a more proactive approach that prioritizes reliability, accountability, and continuous improvement.

It also serves as a reminder that infrastructure policy is not static. The demands placed on our systems are evolving – from increased construction activity to the expansion of energy, telecommunications, and water networks. As complexity grows, so too does the need for systems that can adapt and respond effectively.

Ultimately, the goal is straightforward: fewer incidents, safer communities, and more resilient infrastructure. Achieving that goal requires more than good intentions – it requires clear thinking, disciplined execution, and a willingness to challenge existing practices when they no longer serve their purpose.

The work presented here is a valuable contribution to that effort. It provides both a diagnosis of current challenges and a pathway toward more effective solutions. For policymakers, practitioners, and anyone engaged in the stewardship of infrastructure, it offers insights that are both practical and actionable.

The time has come to ensure that the systems we rely on to protect our infrastructure are as strong and reliable as the infrastructure itself.

Mary served as U.S. Secretary of Transportation from 2006 to 2009. In addition to overseeing over 60,000 employees and annual budgets exceeding $70 billion, she oversaw the implementation of the 811 phone number and program.

Mary E. Peters

Preface

This book exists for one reason – public safety.

Every day across North America, millions of people live, work, and travel above a vast, largely unseen network of buried pipelines, cables, and utilities that quietly sustain modern life. When those systems are damaged, the consequences can be immediate and severe, ranging from injuries and service disruptions to environmental harm and loss of public trust. *Holding Back Disaster* was written to illuminate this hidden world, explain why damage prevention matters, and recognize the people, systems, and organizations working every day to keep communities safe.

The publication of this book was made possible through the generous sharing of insight and support by experts and organizations whose missions are directly aligned with protecting life, property, and critical infrastructure.

Beyond acknowledging that support, this book has a broader purpose – one that aims to serve both the industry and the public.

With a recognition that the industry is navigating a period of heightened debate, competing perspectives, and real tensions, we aim to keep this work centered on the shared goal of public safety and practical solutions, encouraging constructive dialogue without losing sight of the human impacts at stake.

To that end, the book seeks to explain damage prevention to a wide audience, not only how it works but why it matters; to translate a highly technical field into accessible ideas about how the systems beneath our feet are protected; to recognize the industry's practitioners as everyday safety heroes while acknowledging the work still ahead; to demonstrate that damage prevention is a profession with established standards, certifications, and growing professionalization, bringing together a wide range of stakeholders who share responsibility for protecting buried infrastructure; and to appeal to the next generation to enter an industry with urgent workforce needs, meaningful career pathways, and rapidly evolving technology.

Ultimately, our hope is to help make damage prevention a household concept and to raise awareness of the very real stakes involved. We are deeply grateful to the countless professionals across the industry whose leadership and daily work ensure that the systems beneath our feet remain reliable, resilient, and safe. It is our sincere hope that *Holding Back Disaster* contributes, in some small way, to that shared mission.

Scott Landes and Benjamin R. Dierker

Acknowledgments and Board of Reviewers

A special thank you is owed to many in this industry. While we cannot name everyone who has made a significant impact in damage prevention over the decades, we can ensure we thank and recognize those who had a hand in this particular project. From providing sit down interviews to reviewing chapter drafts and more, these dedicated professionals are damage prevention heroes. These are some of the men and women truly *Holding Back Disaster:*

James Anspach, Lawrence Arcand, Dr. Samuel Ariaratnam, Susan Bohl, Mark Bruce, Mike Callan, Glen "Cookie" Cook, Roger Cox, Paul Dickinson, PhD, Mark Drew, Robert Edwards, Mark Frost, Kemp Garcia, Eric Giguere, Mell Greenall, JJ Harrison, Kelley Heinz, Christopher Hill, PE, PhD, CRMP, Sandra Holmes, George Kemp, Bill Kiger, Ellen Kiger, Roger Lipscomb, Brigham McCown, Cliff Meidl, Stephanie Menning, Steven Mohr, John Neilson, Keith Novy, Louis Panzer, Ron Peterson, Meghan Rafinski, Duane Rodgers, Brent Saltzman, Aaron Shavel, PE Jason L. Smith, Bill Stephens, Mike Sullivan, Steve Swazee, Bill Turner, Carl Weimer, Jeff Weise, Sam Wiffen.

The authors gratefully acknowledge the work and assistance of Mr. Colson Grimes, Mr. Albert Bernhardt, Ms. Emma Kelliher, Mr. Minxing Liu, Ms. Nagomi Katano, Mr. Patrick Conrad, Mr. Trevor Mathia, and Mr. Daryoush Jamkhu in helping conduct research on this book.

Additionally, to the many men and women we did not name, despite their great insights, advice, and guidance: **thank you.** Finally, a particular thank you is owed to Mr. Owen Rogers for his tireless assistance in preparing this book.

Abbreviations

811	The nationwide "Call Before You Dig" phone number designated by the FCC
ACTS	Also ACTS Now, which stands for Aligning Chance To Succeed – Publisher of 811 magazines and state 811 conferences that succeeded the Excavation Safety Alliance (ESA) in managing town halls, magazines, and the Global Damage Prevention Summit
APWA	American Public Works Association
ASCE	American Society of Civil Engineers
BEAD	Broadband Equity Access and Deployment Program
CCGA	Canadian Common Ground Alliance
CFR 49 Part 192.614	Federal regulation requiring pipeline operators to maintain damage prevention programs
CGA	Common Ground Alliance, the leading national damage prevention organization
DIRT	Damage Information Reporting Tool, a national database of excavation-related damages managed by CGA
DPAC	Distribution Public Awareness Council, a consortium of natural gas utilities focused on public awareness
ESA	Excavation Safety Alliance, a network for safe excavation and industry education
FCC	Federal Communications Commission
FNCA	Facility Notification Centers Association, a cooperative of One-Call centers across the United States and Canada
NAPUA	North American Private Utility Association
NTSB	National Transportation Safety Board
NUCA	National Utility Contractors Association
NULCA	National Utility Locating Contractors Association
OCSI	One Call Systems International, an early coordinating body of call centers
OCOA	One Calls of America, a cooperative formed to leverage call center efficiency
ODPC	Oceania Damage Prevention Conference
OPS	Office of Pipeline Safety (former name of PHMSA)
PASA	Pipeline Ag Safety Alliance
PCCA	Power and Communication Contractors Association

PHMSA	Pipeline and Hazardous Materials Safety Administration (pronounced "Fim-suh")
PLCA	Pipe Line Contractors Association
SCADA	Supervisory Control and Data Acquisition, a computerized system used to monitor and control utilities
SOCS	Southeastern One Call Systems
SUE	Subsurface Utility Engineering
ULCC / UPROW	Utility Location and Coordination Council / Utilities & Public Rights-of-Way Committee

Introduction

Everything had slowed down and blurred. The atmosphere was thick. Sound slowly returned to everyone's ears, only for sirens to pierce through the haze. One neighbor was just rising up from his knees; others still lay on their lawn with bits of drywall, shingles, and debris on and around them. The front right tire of a silver Nissan Sentra was over the curb and errantly seated on the grass, as the driver white-knuckled and wide-eyed, tried to catch her breath, foot depressed so far onto the brake pedal it nearly reached the floor. Paralyzed in shock, she could not move. Those around the street who could move did so at a confused lethargic pace.

The once still and quiet Thursday morning had become a nightmare of chaos, destruction, and confusion. One house was obliterated from the neighborhood, the two on either side wore open wounds, with entire sections of walls torn down and roofs blown off. First responders had dozens of people to attend to, but there was no hope of saving the Johnsons who lived at 3510 Walker Street.

As police, fire, and ambulances descended on the scene, the road was entirely impassable. Neighbors from way up the road stood on their porches and lawns in disbelief, frozen, unsure if or how to respond. As the police set up a perimeter and spoke with bystanders, fire and medical personnel quickly set up triage tents. Four people were assessed and bandaged up under the first tarp, while two were rushed away in an ambulance.

Later in the morning, when the dust settled and primary findings were recorded by on-scene responders, they would finally remove those at the center of the blast. With deliberate care, they covered Mrs. Johnson's body beneath a sheet – beside her, a smaller one, just as still, just as silent.

A slow trickle of water was running down the street alongside the curb from a water line that was severed in the explosion. Smoke was ascending from the back of the property. Power, internet, and telecommunications services were all disrupted for the immediate neighborhood – 22 homes. The blast had taken out both overhead service drops and buried lateral lines for several houses, including a transformer and one telecom line bringing service into the neighborhood itself.

Walker Street was usually a relatively well-trafficked thoroughfare connecting a main road from the highway to a business park, apartment complex area, and strip of restaurants and businesses. A normal day saw hundreds of cars gliding through. But after 8:25 am on this Thursday morning, no traffic would make it through. Many still sat in their cars, either in shock, with morbid curiosity, or simply unable to maneuver themselves out of the neighborhood yet with all the emergency personnel coming and going.

Those who did see the traffic jam before arriving on scene had a two-mile detour to make it to work or wherever their morning commute took them. The added traffic also slowed police, fire, and medical personnel arriving on scene or taking wounded neighbors to the hospital.

DOI: 10.1201/9781003796619-1

From radio chatter, police interviews, and notes, along with the obvious visual display, anyone arriving on scene would know what happened here. A natural gas leak led to an explosion that leveled the house, sent a fireball across its entire property, and tore into the houses on either side. The immediate effects were clear, but it would take weeks to know more, and years before all the legal matters and financial impacts were resolved.

Nothing could prepare Mr. Johnson for the call he would receive at 10:33 am telling him that there had been an accident at his home, where his wife and child were. The roots of this disaster had been set into motion well before the explosion.

A backhoe operator had gotten up early for landscaping work in a nearby commercial property 300 feet away from the Johnson residence, while temperatures remained cooler. At around 7:07 am, just as he began work, the backhoe operator felt the bucket pull on something hard. Shortly afterwards, he called 911, and then 811 to report that they had struck a pipeline.

A local fire truck arrived on the scene less than 15 minutes later, followed almost immediately by local utility board employees. The responders could smell a strong natural gas odor and discovered water running on the ground.

The backhoe had caught a 2-inch steel natural gas pipeline operating at 51 pounds per square inch, which created a leak at the spot of contact, as well as a second leak where the pipeline had become disconnected from a nearby service line. The backhoe had also damaged a 6-inch water main pipeline, which was spraying water wildly at the damage site.

Shortly after arriving on the scene, the local utility board employees contacted the city to shut the water off in order to access the natural gas leak. The utility employees could see bubbles of gas rising from the now-still water and mud, and began carefully excavating the site to find the exact location and size of the leak.

The utility board employees debated whether the leak was from a gas main, or a smaller gas service line. The owner of the commercial property on which the incident initially occurred shut off the property's gas line.

Just after 8:00 am, a few locals began to congregate in the area, attracted by the now silently whirling fire truck lights and the flood of water, which had left small streams flowing into the street. At 8:14 am, the utility board employees decided to squeeze off the gas main, hoping to stop the flow of gas. As the visible majority of the leaking bubbles ceased, everyone relaxed. The situation seemed under control.

However, unbeknownst to the responders, the leaking colorless natural gas had found its way into the Johnson family basement a few hundred feet away and accumulated. Mr. Johnson didn't notice anything that morning as he rushed to the garage and out of the house early for a meeting. He passed a firetruck on the way to work, but it never occurred to him where it was going. He took his oldest son to school that morning to drop him off for early-season track practice.

At home, his wife and their younger son were getting started with their day. Mrs. Johnson had woken up early to pray and take care of her young son. She spent 20 minutes in her room reading her Bible in the corner chair, before cleaning up and dressing her son. She was putting on his shirt when he became excited at the sound of a nearby siren. He wanted to go see the firetruck, but the siren stopped almost at once, and his mother assured him that the truck was gone.

The silent gas was all the while percolating from the basement up into other corners of the house.

After settling down her son again, Mrs. Johnson made her way to the kitchen. A vague smell of foul eggs wafted into her nostrils, but it was faint as she was recovering her sense of smell still after a cold. What she did perceive, she assumed was the kitchen trash can, some concoction by her boys left out the night before, or maybe even her own imagination. Her haste to prepare for the day ensured she pushed forward.

She pulled some bread from the refrigerator and set it on the counter, then filled a tea kettle at the sink and placed it on the stove top. The water pressure was lower than normal, but she didn't think anything of it. Her head was a bit fuzzy, and she felt the beginning of a headache coming on, but again, thought to herself that it must have been the cold. She placed two slices of bread into the toaster, pushed the lever down, and immediately ignited the stovetop to heat the kettle.

At 10:33 am, Mr. Johnson's phone rang. In one breath, he learned about the explosion that would change his life forever.

The waking nightmare Mr. Johnson seemed destined to enter, mercifully, never came to pass. What you have just read was not a record of events, but a glimpse of what might have been. None of the terrible consequences unfolded. This gut-wrenching tragedy did not occur. It was an avoidable incident – and on this day, it was avoided. As Mr. Johnson took his oldest son to school and headed to work, he would later return home to his wife and youngest for dinner, never aware of how close disaster had crept. Today, the Johnsons lived a normal day, spared by the unseen diligence of a young man they would never meet. Days earlier, a utility locator had marked the buried lines, quietly preventing this and countless other tragedies.

Jacob Albright's eyes were heavy as he depressed the button on his alarm at 4:50 am that Tuesday. He was exhausted from being out late (until 10:30 pm) because his brother was in town over the long weekend. The two hadn't caught up in a few years. Both military veterans, they were bonded in deep and unexplainable ways, from their youth and family connection to their later-found unity as brothers in arms.

Jacob always made time for his brother. But he is also a working man and has responsibility. This morning, it was to wake up before the sun, and get out to the job sites.

After brewing his coffee, filling his go-cup, and grabbing his vest and hard hat, Jacob jumped in his truck and rolled out. He had 24 houses, lots, and ranches to visit today – and they were not close together. He knew he would be darting across the county today, so the early start was essential.

To avoid alarming people or having the cops called on him, Jacob started where there were no people to be found. Someone was planning to dig on an empty lot near a section of rural highway. This would set him up to hit a ranch, where a farmer intended to build a new fence by driving long metal stakes into the earth. The second part of his morning would be all residential. Working with his project managers and the administrative team at his locating company, he'd formed a perfectly optimized loop that would finish all his work and bring him right back home.

As he drove from job to job, the small colored flags in the bed of his truck dwindled in number and a few empty cans of spray paint started to pile up in a plastic bucket. Besides what was laid down on the job sites and the state of his truck bed, there was evidence of his work in colored mist traversing his boots. Today, orange is a very popular color.

This book tells the story of the world beneath your feet, the men and women making the unseen visible, the critical systems that serve you every day and how they are constantly at risk, and how it has all evolved to keep you safe. Moreover, it serves as an invitation to participate.

Whether it is simply by engaging with the 811 system and being more aware of your surroundings, or joining the ranks of this profession and helping shape its future, you have a role to play.

Damage Prevention – a phrase you'll see innumerable times within this book, is the name for the industry of people helping prevent harm to critical infrastructure, primarily harm from excavation work or digging. While overhead lines present their own risks and costs, damage prevention primarily focuses on keeping pipelines, cables, and wires buried underground safe. They are uniquely at risk because tens of millions of miles of them across the country are hidden from view by the dirt. That means every ground disturbance creates a risk of striking a pipe, cable, or wire no matter where you are in America and regardless of whether you're a construction crew or do-it-yourselfer at home.

An equal focus and alternative name for the industry is "excavation safety," because virtually all damage comes at the tip of a shovel or power tool, and keeping people safe is a shared responsibility. Excavators, again, can be professional crews or even homeowners doing some landscaping for the weekend. *Damage Prevention* refers to protecting the infrastructure and *Excavation Safety* refers to protecting the worker, but these are two strands forming the double-helix of the same DNA.

Communication and collaboration between and among many different people and sectors make damage prevention possible. From homeowners and excavators planning to dig into the ground to anybody who owns or operates pipes, cables, and wires under the surface, locate technicians spray painting the surface to 811 center operators keeping everyone looped together, and even public agencies, regulators, and more – the damage prevention community is diverse and wide ranging.

The industry is nearly a century old, but it is increasingly a hotbed of technology and innovation. Like haircuts or car washes, it is an evergreen industry that will always be needed. Every day, more and more infrastructure is built to support our growing population, and as new needs arise, new services must connect utilities to consumers. This demand requires damage prevention roles at utility companies, transportation departments, technology companies, construction firms, and beyond.

Not only is the industry growing, but it is being mapped out more clearly so that new entrants can visualize the whole ecosystem. It is being more professionalized to provide for career stability, trajectory, and dynamism. And above all, the damage prevention industry is ready for its next phase, poised to deploy new technology, define new best practices, and certify excellent workers as comprehensive damage prevention professionals.

In the pages that follow, you will get to join the professionals keeping you safe and protecting the vital services provided by the infrastructure, which your communities rely on every day. Each chapter begins with a short, fictionalized account that drops you into the shoes of a different role – locators, excavators, utility crews, and other stakeholders – going about a familiar part of their job or providing a visual to thematically frame the chapter. Whether you're new to the industry or a seasoned veteran, you'll see through the eyes of dozens of individuals whose lives and work intersect beneath the surface – dozens of voices, roles, and responsibilities that make up the living, breathing system of damage prevention.

But you won't just be a passive reader, you will jump straight into the industry by rubbing elbows with all the different stakeholders and hearing directly from them. This book is intended to be immersive, like attending a conference yourself. If you're an old damage prevention veteran, consider this your next conference (and we hope you appreciate the reduced price of entry!). If you're brand new, this is your rite of passage – your first conference – where you'll meet the key players for yourself by attending virtual sessions, each packed with insight from decades of experience and multiple perspectives. You'll also see additional resources at the end of each chapter. These aren't dusty old footnotes, but live resources like videos, mobile apps, reports, and organizations to look into as you learn more.

Welcome to a conference in a book! Not only will you hear from the authors, but you are holding an all-access pass to visit every session. You won't miss concurrent talks held in different rooms, because they are all laid out here. The authors are your hosts; the quotes you'll encounter throughout the book are real insights shared by the real men and women *holding back disaster*. And like at a conference, you can think of these quotes as session comments, hallway conversation, exhibit pitches, and even on-site podcast interviews where incredible insights and wisdom is regularly shared. You may also read the same information presented in multiple chapters, but explained from a slightly different angle. This is because each stakeholder sees their piece of the damage prevention puzzle differently, and like at a conference, every session will repeat or explain certain concepts to make sure the given audience is up to speed.

We come to this industry from two different vantage points – Scott from decades in publishing, conferencing, designing/manufacturing damage prevention systems, and more; Benjamin from legal and public policy research at the nation's only think tank dedicated to infrastructure. The shared goal of this work is to bring Damage Prevention into the light, elevate it for the world to see how it enables our daily lives to function, and commend the men and women making it successful. From there, to invite the next generation to shape and carry it on!

Chapter 1

Getting Your Bearings

It was well past midnight, and the Speaker of the House was exhausted, as were his top staffers and most people on Capitol Hill. The Majority Whip was up well into the night ensuring they would have all the votes they needed on the major infrastructure package. There had been talk for years – decades even – about investing in American roads, bridges, dams, energy grid, water systems, and more. Finally, the stars had aligned (although it felt more like they had been wrangled and dragged into place).

Both Republicans and Democrats came together to approve the mega legislation that would jolt tens of billions of dollars into infrastructure revitalization. From programs to expand broadband access to new highway projects, pipe replacements, and more, building would be back on the menu for American workers. There was a palpable atmosphere in the U.S. House of Representatives as the final vote came up. So many people had worked so hard to hammer out deals, make the math work, and ensure there were enough votes.

When the vote passed the majority threshold, cheers broke out in the chamber. Representative Stevenson shouted "Huzzah" and looked around for colleagues, shaking a celebratory fist in the air. The budget hawks in the room had looks of either disgust or simple defeat. They saw the legislation only as a giant debt balloon for their children and grandchildren. Others saw it as an investment with real return that would juice the economy.

In the back of the room, a 30-year-old policy analyst on the Transportation and Infrastructure Committee looked like a deer in the headlights. He had worked on the legislation for months and taken hundreds of meetings with lobbyists, nonprofits, construction firms, and more. He may have been the best-informed person in the entire room. His take differed from everyone else's.

When Jack Villaluz saw the vote, he wondered if anyone else in the room knew what they had just done.

Jack was a relatively junior staffer, but he had advanced degrees in economics and business. He spent a few years working for Ernst and Young and his entire portfolio focused on construction and infrastructure topics. He learned a lot and wanted to bring that into a public service career, so he pivoted to the government, coming to work for the U.S. House of Representatives.

What Jack saw – that no one else seemed to see – was that every dollar and every well-intentioned project Congress intended to help rebuild America could also create economic challenges they never intended. When he thought of money pouring into a construction site, he visualized large machinery breaking ground, with workers so eager to get to work, projects so long delayed, and economic potential so ripe for the picking, that corners were cut,

DOI: 10.1201/9781003796619-2

and miscommunication occurred. He saw all that heavy machinery ripping up concrete and bringing brand new fiber optic cables up with it. He saw workers digging a trench and the walls collapsing in on them. He saw traffic backed up for miles because a road project struck a natural gas pipeline, and an entire intersection was charred as emergency responders triaged the scene.

Jack was not a pessimist. But he could see a step or two ahead. He hoped the funds and projects approved would be good for local communities and the national economy. But he knew there was at least a chance that a lot of things would go wrong along the way. Jack's unease lingered long after the celebration ended. The nation saw a windfall; he saw the hidden costs beneath the surface.

When Heath Ledger's Joker set a pyramid of cash ablaze in The Dark Knight, in-movie characters and audiences alike felt a pit in their stomachs. A mixture of rage, confusion, and righteous indignation over the waste washed over onlookers. *How senseless! How much could be done with those precious resources?*

But in the end, viewers know it was all fiction. The billion dollars destroyed amounted to nothing more than tension and angst that were necessary to the plot and forgotten as soon as the movie concluded. No one's day was ruined at the thought of a fictional chaotic villain wasting mobsters' cash.

One wonders how these same people – momentarily and situationally aghast for a single movie scene – might react to a real-world Joker's inferno. Surely a sum so carelessly destroyed would make them sick, if not enraged.

While it may be hard to imagine, a far greater sum is regularly set ablaze in the United States – and while the flames may be metaphorical, the cash is every bit as lost.

For fiction to catch up with reality, the real-life Joker would need to commit his same arson every other day continuously without relenting; and repeat it every year. Even then, he may not manage to match the economic harm plaguing American infrastructure, construction, and communities.

The Scale of the Problem

Every year in the United States, tens of billions of dollars are lost and wasted within just one industry. It is the result of excavation damage, the system designed to prevent it, and the economic ripples reverberating through communities when damage occurs. This multi-billion-dollar annual economic harm is a combination of direct damage to pipelines, buried utility lines, and subsurface infrastructure, alongside indirect costs that come up while resolving the damage or that impact communities nearby. A host of entirely separate losses, distinct from damage costs, include billions of dollars forfeit to inefficiency and waste within the system trying to prevent the damage in the first place. While many stakeholders may disagree with the specific inefficiencies, there is little doubt that the scale is vast, and when combined with damage costs, is considerable.

It does not have to be this way. Rather than helplessly watch our economic resources burn up in chaotic infernos, we can reform key elements of the system to cut costs, add value, and above all, prevent damage and keep excavators safe.

Flattened home resulting from a gas leak.

Image from Shutterstock.

Beneath the Surface

To understand the issue and the stakes at play, first think about the world's leading economy in the 21st century. The United States is a dynamic powerhouse of goods and services, energy and information, and movement of commodities and waste. With a population nearing 350 million people, it takes immense resources to keep the lights on, power the economic engine, and keep modern life moving.

Readers would of course recognize the roads and bridges, the overhead transmission and distribution power lines, and even cell towers. But so much of the infrastructure that we rely on is beneath our feet. Some utilities pop up only minimally like a meerkat peeking up from its subterranean lair – think of a fire hydrant, clearly the iceberg tip of a mainly underground system. Others remain almost entirely subsurface, like fiber optic cables buried within conduits, out of sight and out of mind.

All told, tens of millions of miles worth of pipes, cables, and wires crisscross the nation just underneath the surface. These include natural gas and hazardous liquid pipelines, water lines, internet and telecommunications cables, electrical wires, sewage mains, and dozens of other facilities. These are owned or operated by municipalities, private companies, community organizations, and even public agencies. They keep energy, information, and resources running into our modern buildings, vehicles, devices, and more, while taking waste and other content out.

With the high concentration of utilities hidden by the opaque surface of the ground, it is all too common to hit them when digging. That is true of home projects with a shovel, trowel, or post hole digger and it is true of construction sites with power augers, jack hammers, back hoes, and other excavation machinery. Striking these pipelines and buried utilities inevitably results in damage.

What Is the Damage Prevention System?

Considering the constant excavation projects across the country taking place every day, the need for an organized system to avoid damages is clear. To reduce the risk of striking any buried infrastructure, there is a simple and standardized process in place in every state. Before any ground is disturbed, by law the excavator must call 811 or submit an online marking request (or "ticket") for identifying and locating any subsurface utilities. This notice is fielded by a statewide or regional center (historically known as a "One-Call Center" now known by the shorthand "811 Center"), which then alerts the relevant utility companies with facilities in the area. In turn, those companies dispatch professional locators to identify and mark the approximate location of underground lines using paint, flags, or stakes. This service is free to most excavators. Once the site is properly marked and the legal timeframe is completed, the excavator can begin digging with a better understanding of what lies below and how to avoid it.

Throughout this book, we will use the terms 811 Centers, One-Call Centers, and Facility Notification Centers. In recognition of industry trends and technology shifts, we will use "811 Center" when discussing U.S. organizations and "Facility Notification Centers" to refer to this same group while including international examples, most notably Canada and Australia. Finally, "One-Call Center" will be used when context demands, or historical documentation requires it to avoid anachronism.

With so many millions of miles of buried pipe, cable, and wire, digging anywhere is a risk. And excavators may not know what type of lines are on a site or who owns them, so contacting the 811 Center every time is paramount. Industry research indicates that every year, tens of millions of Americans will dig without following the full process. In fact, in 2023, the Common Ground Alliance (CGA) explained that a recent, "Survey reveals 49.3 million Americans plan to dig without contacting 811 first, risking disruption to critical services."[1] That means nearly 50 million people

> ...will put a shovel, backhoe, excavator, trencher, auger or other tool into the ground this year without getting buried lines located first. That's [nearly 50 million] chances for a gas explosion, a petroleum products spill, cutting off water to a hospital – it's all happened before.[2]

Industry leadership goes on to demonstrate that scale and potential impact.

> "Four in 10 U.S. homeowners who are planning to dig on their property this year will not contact 811 at least a few days in advance. That's more than 49 million Americans who are putting themselves at serious risk of personal injury and utility service interruptions," said CGA president and CEO Sarah K. Magruder Lyle.[3]

Although millions of people may have interactions with the damage prevention industry, there are three core stakeholder groups to note at the start. *Excavators* are heavily involved in the damage prevention industry because most underground utilities are damaged during excavation projects. Although homeowners and landscapers often notify the 811 Center system, the vast majority of requests to locate buried utilities are submitted by professional excavators. The *811 Centers* field locate requests from excavators and utilize mapping to determine which utility companies are affected in the area of excavation and notify them accordingly. However, it is not the 811 Centers themselves that mark underground facilities. Instead, that responsibility falls to professional locators, who mark the ground using the flags and paint familiar to us. *Locators* may be directly employed by a utility owner, but often belong to professional locating firms contracted by utility owners.

The 811 Center process is designed to identify and mark public utilities, which may be less than half the total buried facility mileage. The remaining portion includes private lines of all kinds, which may be "behind the meter" or include things like sprinkler and irrigation lines not associated with public utilities. While an entire system primarily focused to locate and protect only half of all buried infrastructure may seem disproportionate and ineffective, public lines represent the most dangerous and economically significant (e.g., natural gas lines, etc.). The 811 Centers also regularly educate and inform callers about private lines, making sure these do not go unaddressed, merely that they require a private locator and are not located and marked as part of the main 811 process. This type of education and the work 811 Centers do on a daily basis is foundational to damage prevention, precisely because they interface with so many stakeholders at once.

> The responsibility for preventing excavation damage is shared by all stakeholders, and includes elements such as planning, effective use of one-call systems, accurate location and marking of underground facilities, adherence to safe digging practices, proper placement of facilities, and strong public education and awareness.
>
> (Common Ground Study, 1999)[4]

The well-established 811 process is relatively straightforward. Yet, many still do not know about it or do not fully follow each of the right steps. Missteps in this process – whether intentional or accidental – carry serious risks.

When Things Go Wrong

Recalling the Joker once more, here is where losses are both financial and flammable. Unfortunately, natural gas explosions are not rare enough. Nationally, they are the second-most struck facility.[5] Unlike the Joker's stunt, these explosions aren't fiction.

When these lines are damaged, volatile methane leaks from the pipe, and a simple spark from the machinery can ignite the gas. Even worse outcomes occur when a slow leak has been undetected for hours or days, and an ignition leads to an explosion that can level an entire home or even a neighborhood. Needless to say, these accidents often result in serious injury or death – enormous costs in themselves – in addition to physical damage to buildings, machinery, the pipelines and surrounding infrastructure, lost product (e.g., the gas), and significant delays for the construction job. Added to these are the costs for emergency response and medical care. Roads are closed or traffic rerouted, resulting in delays, lost productivity, and economic disruption for local communities. There may be reputational harm that hurts the business of local utilities, the excavator, or even locators who may have been involved in the project. This can lead to economic losses even before accounting for the deadweight loss of litigation to pursue ultimate liability for the damage itself. Additional service disruptions like power loss and internet outages are common when major damage events occur, adding once more to the rippling economic consequences of a single incident.

Repairing communications cables can be extremely expensive and leave people without vital services for extended periods.

Rippling costs are referred to as the indirect or societal costs (i.e., those outside of direct damage, product loss, and repair costs), and economic literature suggests that a standard multiplier is 1:30 for direct to indirect costs.[6,7] That means if a natural gas explosion resulted in $1 million in direct harm, the total cost is a combined $31 million for the local community as the shockwave of damage, repair, clean up, disruption, delay, liability, and other effects pass through the region. Knowing just how many incidents occur and how much they cost is a difficult task. And as you'll see, determining the extent of the impact relies on total damage numbers, cost estimates, system efficiency considerations, economic factors, and more.

The Data Dilemma

The CGA took on this challenge directly and began compiling damage reports and modeling total estimated damages over 20 years ago. Their work relies on robust stakeholder engagement, as it is a voluntary reporting system. That means when an excavator hits a line, they may choose to notify CGA or may stay silent. Perhaps a locator or the 811 Center is notified and they can submit the incident to CGA. The utility company itself, when it finds out, may report. In a perfect world, all these parties would report the damage, and CGA would have many reporting sources for the same incident, which they are adept at filtering through and identifying both total reported damage number and unique damages, when they align multiple reports of the same incident.

However, with voluntary reporting, exact certainty is elusive. That is why CGA has developed many models over the years to estimate total damages, using confidence intervals and upper and lower bounds. Over time, changes in reporting levels and lack of sufficient data from voluntary stakeholder reporting to the Damage Information Reporting Tool (DIRT) managed by CGA has necessitated a totally new approach. In 2024, the model changed again to utilize a base-year ratio that does not estimate total damages but a percentage increase or decrease from the baseline year. This uses a formula to estimate damage numbers using reliably consistent reporters over the time frames as well as specific outside data points.

There is good data for certain damages, however. Incidents must be reported to the U.S. Department of Transportation's Pipeline and Hazardous Materials Safety Administration (PHMSA) when they involve hazardous material and pipeline strikes. Yet others like telecom or water lines are under no federal reporting rules, and often fall in the gaps of state laws as well. Then there are simple compliance issues, where smaller damages – even when reporting to some authority is required – go unreported. State laws vary widely here, and where CGA is concerned, few state laws compel DIRT participation.[8]

Regardless of who collects the data and what modeling is employed, it is extremely difficult to get an accurate picture of total damages. CGA is to be praised for creating a reporting and display platform that is among the best in the world.

Some stakeholders believe most excavation incidents go unreported. A good starting point is to look at the last available estimates from CGA in its reporting from 2019 to 2023. In those years, based on only reported damages and extrapolating through statistical modeling, the reports point to over half a million annual damages. While the models are intended to present estimates of total damage, the extent of unreported damages is virtually impossible to know. Subsequent model changes have made year-to-year comparisons difficult, but CGA noted a marginal decline in estimated damages in 2023 and a marginal increase in 2024.

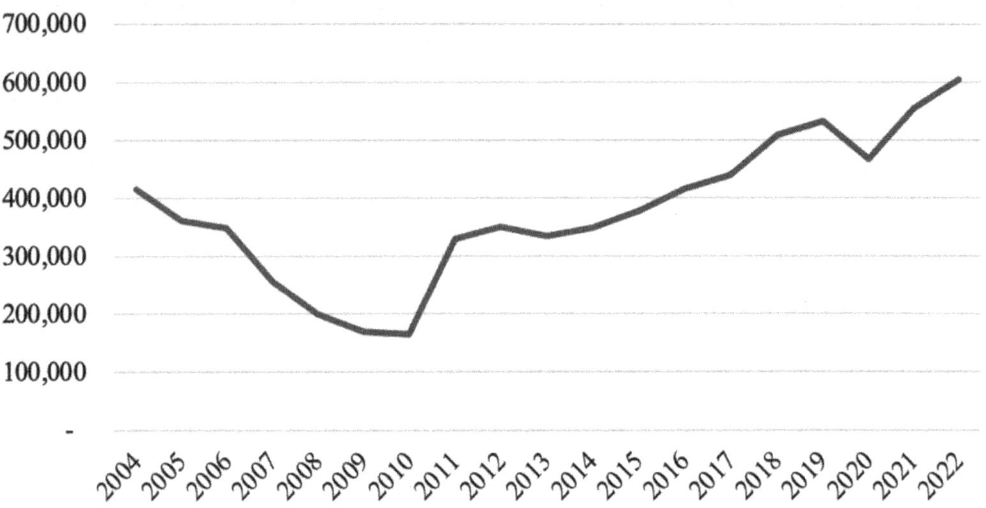

Total estimated excavation damage trend over time.

Tracking the Damage

What we can say with confidence is that, in 2019 for instance, more than 532,000 damages took place costing more than $30 billion to the United States that year. This is a conservative case outlined by CGA in its annual DIRT Report. Many believe the true damage numbers and costs are much higher. We also know damages increased in subsequent years – which would increase both the number of incidents and the resulting costs.

From there, we know that construction spending is highly correlated with damage. This makes intuitive sense, because new construction projects involve breaking ground, and the more that occurs, the more opportunity there is for a strike or near miss – even when the right procedures are followed, which is never a guarantee. Trends demonstrate enormous federal spending into the construction sector in recent years, including from the Infrastructure Investments and Jobs Act – also commonly referred to as the Bipartisan Infrastructure Law – and enormous advancement in fiber optic installations, among other national, state, local, and private initiatives.

CGA noted that 2019 was at the time a record high year for excavation damage. While 2020 was an anomalous year, with COVID-19 impacting all industries, data suggests a slight decrease in damages. In 2021, CGA reported an estimated three-year trend increase. That would make it a new all-time record year. The following year, a similar marginal increase was reported. With government and private sector construction spending increasing and no clear or discernable reason to believe there was improved adherence to excavation safety procedures, we believe new records in excavation damage have occurred in recent years that push the total estimated damage number closer to one million incidents by 2025. Indeed, the latest DIRT Report released in 2025 demonstrates another uptick in damages after the prior year introduced a new model that showed a marginal decline.

Model changes at CGA make this difficult to cross reference, as many voluntary reporting sources disappeared from the DIRT platform several years ago. The result has been DIRT Reports without estimated total damages and new models with significantly less data available to make strong statistical claims about the true total damage numbers each year.

To be conservative, we will say that every year in the United States somewhere between 600,000 and 1,200,000 incidents take place in which an excavator strikes a buried pipe, cable, or wire.

With this level of damage, what are the costs today? We can start again with the reputable estimation from CGA in 2019. For that year, CGA estimated the economic cost was $30 billion. CGA explained in its 2019 report that the true cost could have ranged from as low as $12 billion to as high as $60 billion. The $30 billion figure was a median value, but our review of the data sources indicates the higher value was likely a closer reflection of the cost at the time within the United States. Taking the true cost to be closer to $60 billion then ensures our ultimate estimate for today is most conservative.

We then must adjust for the increase in total damages since 2019. Not only do we know that over the last six years, excavation damage incidents increased, but there has been considerable inflation. Together these push the annual cost of excavation damage far higher. Finally, more recent studies have pushed the social cost multiplier higher than 30:1 for excavation damages on certain utilities, meaning that in reality, more costs are accruing from incidents due to urban population density, interconnected infrastructure, and more.[9]

To be conservative, we adjust for inflation and use data from CGA DIRT Reports to estimate that in 2025, annual excavation damages cost the United States over $75 billion.

Systemic Waste and Inefficiency

Finally, there are costs to operate the system and the inefficiencies within it. While we cannot count the general operational costs as losses – especially if they result in preventing damage, thereby fulfilling their purposes and sparing costs – we can count those outdated practices still employed or activities not associated with preventing more damage. According to a coalition within the industry, some common costs include:

> …utilities and third-party locators needlessly sent to locate lines for construction projects that then do not happen; poor instructions given to locators, causing wasted time or additional work; locate marks destroyed by construction and then requiring reinstallation and contractor wait time when location efforts exceed the legal notice period.[10]

The group espousing these findings is known as the Infrastructure Protection Coalition (IPC), which came on the scene only in recent years. It is a coalition of several contractor associations. While the coalition came somewhat out of the blue for many in the industry, their formation was to advance a data-based report filled with statistics and was well-sourced with hundreds of citations. It certainly caused conversation within the industry, in part due to the title of its report: *811 Emergency.*

While the rhetoric was strong and perhaps hyperbolic, the underlying data was grounded in reality and can be relied upon for helping understand the broad strokes of costs within the industry. In IPC's nationwide report, they identified a total of $61 billion in total waste and inefficiency costs in 2021. Intended to represent annual waste, this means that every year, damage prevention processes may be needlessly adding to economic harm while merely going through the motions of the damage prevention process that could be tightened up and refined.

It is reasonable to believe some of these costs were addressed in the time since the report. We can assume this on the grounds that some stakeholders likely reacted to the report itself and adopted some of its recommendations, and because new technology and practices have been further honed since then. But the scale of costs is undoubtedly still intact. We can know this by general stakeholder consensus that no major seismic shifts or disruptions have occurred in the last five years, and because state laws are slow to change, and little legislative or regulatory change has been proposed or taken effect since.

To be conservative, the inflation adjusted data from the IPC report indicates that in 2025, inefficiencies in the system cost the economy over $75 billion.

What all of this means is that today, damage numbers are high, and costs are higher. Conservatively, the annual costs and losses associated with damage to buried infrastructure likely exceed $150 billion. That is enough to purchase 12 new aircraft carriers. Alternatively, this sum could power 90 million U.S. homes with electricity for a whole year, fully two-thirds of U.S. households. In less than 10 years with no reform, over a trillion and a half dollars is vanished from the productive economy. Beyond damages, industry stakeholders are under strain.

> If you're in the One-Call center, you're trying to figure out how to address volume increases with staffing or technology. If you're in the locate industry, you're trying to get a handle on that sheer volume, the wave of tickets that are coming and trying to figure that out. If you're a contractor who's doing this work, you've got incentives on your back to try to get this work done as quickly as you can.
>
> (Louis Panzer, Executive Director North Carolina 811, *ESA town hall*)

However, with hard work and communication, these damages are avoidable, and even reversible.

A Path Forward

There is reason for optimism, and that is what this book spotlights.

Working to prevent all the losses described above are the professionals within the industry of Damage Prevention. These key players and dynamics are described in the following chapters to help tour readers through the ecosystem. They are the locators who are responsible for identifying and making surface markings of where buried utilities are before any excavation work begins. Excavators are the ones digging and the last line of defense to ensure infrastructure is not disrupted, whether they are installing new facilities, maintaining existing ones, or simply digging for another purpose. Utility companies are central figures, both because their assets are the things at risk, but also because they have the most detailed information about the presence and location of their own buried lines. The 811 Centers field calls and online requests from the excavators and notify the utility companies to send out locators. Nevertheless, utility companies are at the core of every process: they or their contractors perform roughly 70% of all excavation, they hold the contracts for locators, and they typically fund 100% of the 811 Centers.

Beyond this core set of stakeholders are many others. As CGA counts it, there are 16 stakeholder groups. Representing cross sections of both public and private sectors, these also include things like emergency response, state transportation departments, and regulators. Unwinding all the intersecting lines is the project of this book, not only to demonstrate how complicated it can all be, but to show how much room there is for collaboration and what options exist for career trajectory and mobility. The more intertwined all the stakeholder groups are, the more potential a young damage prevention professional has to change positions throughout his career, gathering skills like a snowball, and ultimately becoming an industry-wide certified professional.

> It's an evolving practice, and damage prevention, it never really ends. We're always improving our methods.
>
> (Mike Sullivan, President, Utility Safety Partners, *ESA town hall*)

Extinguishing the Inferno

With those skills and a bit of ambition, the rising professionals in the damage prevention industry will be ready to squash any Jokers that come and the costs they may seek to inflict on local communities and the national economy as a whole. Together, we can extinguish the entire conflagration and restore peace and security to homes, workplaces, and balance to the ecosystem.

Chapter 1 Resources

The free library of over 40 Excavation Safety Town Halls provides insights and education on key damage prevention topics.

CGA – A member-driven association of nearly 4,000 members across the entire underground utility industry.

DIRT Platform – An online interactive dashboard providing comprehensive accounting and analysis of excavation-related damages to buried infrastructure in the United States and Canada.

Chaos to Common Ground in 40 Years – A 2014 presentation at Infrastructure Resources CGA Excavation Safety Conference & Expo.

Infrastructure Protection Coalition – 811 Emergency Report.

Notes

1 Common Ground Alliance (CGA). (2023, March 29). Survey reveals 49.3 million Americans plan to dig without contacting 811 first, risking disruption to critical services. https://commongroundalliance.com/Publications-Media/Press-Releases/survey-reveals-493-million-americans-plan-to-dig-without-contacting-811-first-risking-disruption-to-critical-services
2 Underground Infrastructure. (2017, July). 3 magic numbers for underground construction, maintenance. https://undergroundinfrastructure.com/magazine/2017/july-2017-vol-72-no-7/general/3-magic-numbers-for-underground-construction-maintenance
3 Common Ground Alliance (CGA). (2023, March 29). Survey reveals 49.3 million Americans plan to dig without contacting 811 first, risking disruption to critical services. https://commongroundalliance.com/Publications-Media/Press-Releases/survey-reveals-493-million-americans-plan-to-dig-without-contacting-811-first-risking-disruption-to-critical-services

4 United States Department of Transportation; Research and Special Programs Administration; Office of Pipeline Safety. (1999, August). Common ground study of one-call systems and Damage Prevention Best Practices. https://primis.phmsa.dot.gov/comm/publications/CommonGroundStudy090499.pdf
5 Common Ground Alliance (CGA). (2025, August). 2024 DIRT Report. https://dirt.commongroundalliance.com/Portals/9/Common-Ground-Alliance-DIRT-Report-2024.pdf
6 Makana, L. O., Metje, N., Jefferson, I., Sackey, M., & Rogers, C. D. (2020). Cost estimation of utility strikes: Towards proactive management of Street Works. *Infrastructure Asset Management*, *7*(2), 64–76. https://doi.org/10.1680/jinam.17.00033
7 Bradbury, M. (2024, November). Economic assessment of utility strikes in Australia. https://www.aihs.org.au/common/Uploaded%20files/Learning-and-Events/BYDA%20Economic%20Assessment%20of%20Utility%20Strikes%20in%20Australia%202024_Final.pdf
8 Dierker, B. (2024, July). Excavation Damage Reporting: The Basics of Data Collection and How States Require Strikes and Near Misses to Be Reported. Alliance for Innovation and Infrastructure. https://www.aii.org/wp-content/uploads/2024/07/Excavation-Damage-Reporting-Overview.pdf
9 Bradbury, M. (2024, November). Economic assessment of utility strikes in Australia. https://www.aihs.org.au/common/Uploaded%20files/Learning-and-Events/BYDA%20Economic%20Assessment%20of%20Utility%20Strikes%20in%20Australia%202024_Final.pdf
10 Infrastructure Protection Coalition (IPC). (2022, June 15). 811 Emergency. https://ipcweb.org/images/reports/US-RPT.pdf

Chapter 2

Locating

The Front Line of Damage Prevention

The truck entered the subdivision and wove its way past Wonder Dr, Warner St, Waller Rd, and through nearly a dozen other alliterative street names until the GPS pinged to indicate reaching the destination. The street was curved, and no parking was available in front of the specific job site, so the man parked 100 yards up the road in a small carve out. It wasn't quite a parking lot – it was after all in the middle of a residential street – but it was near a set of mailboxes for a nearby apartment complex and was likely built to give the mail truck space to pull up without blocking the road. It was a very forward-thinking design, and clearly reflected the intentionality of the planning that went into this community that was only built in the last 10 years.

Closing the drivers' side door, Jorge pulled an orange vest and hard hat from the back seat before leaving the truck behind and walking over to look at the job site. The clouds forming overhead signaled a sense of urgency, because there were many other sites to visit today. His need to work quickly was only outweighed by the need for accuracy, because in less than 24 hours, crews would descend on this neighborhood with concrete saws and jackhammers to open the roadway.

He paced by nearly 10 homes before the address number matched what was on his ticket. To anyone else, this would just be a residential neighborhood – a collection of single-family homes with nice green lawns, modest sedans and minivans spread across driveways, and a small dog barking from a backyard somewhere. But that isn't all he saw.

His eyes were drawn to the street lamps; the storm drain opening; a manhole cover. He scanned the side yards for the forest-green utility box that wanted to fade away into the green grass, but stood out like a sore thumb to him. He also took note of what he didn't see. There were no power lines or service cables coming from overhead and reaching out for the sides of houses. He knew what all of this meant. In his mind's eye, he saw the intricate network of pipes, cables, and wires beneath his feet connecting all these homes to the water, power, information, and waste systems that they depend on but which are concealed out of sight.

Jorge knew that out of sight meant risk. If he didn't do his job perfectly, and communicate clearly, someone later in the process could rip into these buried lines and disrupt the entire neighborhood or even lead to casualties. Working swiftly to beat the weather, balancing prioritization of other jobs with technical accuracy at each site, and overcoming signal interference and other distractions, he was focused on each line below the ground.

His job today was to make them known.

DOI: 10.1201/9781003796619-3

Locating Is a Highly Skilled, Vital Profession

Every excavation project, from backyard landscaping done by the homeowner to massive infrastructure renewal projects done by a professional excavation company, begins with a single question: *what lies beneath?* Or as the Canadians might say: *where is the line?*

The *Locate* industry is a stakeholder group within the damage prevention ecosystem. Their job is, as the name implies, to locate underground facilities. Most often utilizing handheld devices and maps, they detect and then make surface-level markings to represent the pathway and type of buried infrastructure. When handheld devices do not have the capability to determine the location of the facility, other tools and methods may be required. This is critical to helping prevent excavation damage incidents, because it helps give excavators a visual representation of what is below.

In some cases, excavators have "as-built" drawings from the facility owners, or even detailed subsurface utility engineering (SUE) plans, which we will cover in detail in Chapter 5. However, the real answer to that critical first question lies in the hands of a locator. Locate technicians are the highly skilled, often underappreciated professionals who find and mark the approximate location of the buried utilities in the area in order to avoid a potential damage or even a disaster. If they make a mistake, miscommunicate, or fail to mark a line, the likelihood that an excavator strikes it goes up dramatically.

Once all facility owners have responded to the locate request and marked any facilities they have in the proposed digging area, which they must do within two to three days as described in state law, excavators are nearly ready to dig. They may begin digging once this process has played out and all communications are cleared up between parties, discussed throughout future chapters. The excavator will arrive on scene, in most cases seeing markings left by the locator.

If the digging will take place outside of the public right-of-way the excavator must contact a private locate company to locate all buried facilities. These facilities can include electric cables, sprinkler systems, propane pipelines, and more. Industry estimates state that up to 65 percent of buried facilities are privately owned and on private property, so if the dig site is on private property it is critical for excavators to get these facilities located before digging.

The locator in the United States will use a standardized color code[1] of spray paint – or in some cases stakes or small colored flags – to identify:

Natural Gas in *Yellow* – which also represents oil, steam, and other gases
Electric Lines in *Red* – which can also be other power, cables, and conduit
Telecommunication Lines in *Orange* – which can also be other cables, alarms, broadband, and signal wires
Potable Water in *Blue* – which may link to fresh drinking water faucets, fire hydrants, and more
Sewer Mains in *Green* – which also include storm drains and waste water lines
Reclaimed Water in *Purple* – often associated with irrigation or non-drinking water

Other standardized colors like pink and white exist, but these are not used by utility locators and are covered later on. Different colors also exist in jurisdictions outside the United States, which is not covered in this book, but is important to note. The basic task of the locator is demonstrating for the excavator what is below the surface and where, by generally

Spray paint markings across the streets and sidewalks of Washington, D.C.

revealing the type and route on the surface. The colored spray paint accomplishes both of these ends.

However, a tolerance zone must be considered with each set of marks. The tolerance zone is the area where the excavator must be careful, because they may still strike a line or have a near miss when using large machinery. Inside of the tolerance zone, which you may think of as a simple buffer for safety sake, or an allowance for lack of precision of the tools and methods, the excavator must hand dig or use alternate excavation techniques.

Tolerance zones vary by state, but range from 18″ to 36″ on each side of the outer edge of the pipeline or cable, where digging with mechanized equipment is not allowed.[2] As depicted in the following diagram, if the tolerance zone is 24″ and the pipeline diameter is 2″ that translates into a 50″ area where no mechanized equipment may be used.

While helpful for safety, tolerance zones do not account for all scenarios. Locate markings do not provide depth information, and not all utility lines are grouped together in a conduit or layered tightly in a utility corridor. Many utilities could be at different depths and essentially side by side.

What is often overlooked or taken for granted by some is that even if everyone does the right thing by requesting locates, waiting the proper time, respecting the marks, and digging by hand in the tolerance zone, if the locator mislocated the facilities, the odds of buried facilities being damaged increase dramatically. Locating buried facilities accurately requires thorough training, the right equipment, and good communication between the facility owner, the locator, and the excavator. Perhaps most importantly, it requires enough time to do the job properly.

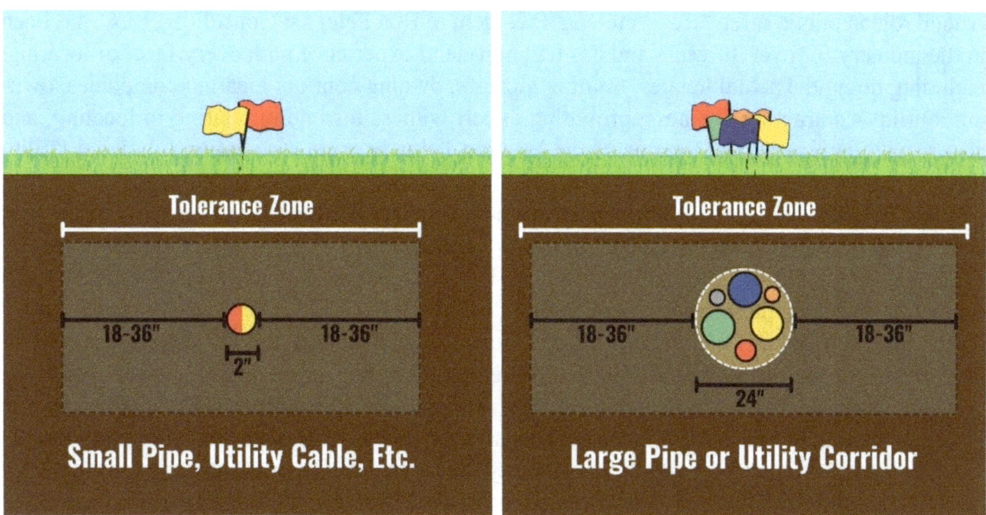

Representation of how a tolerance zone extends to each side of a utility.

Heroes Without Capes

Locators are essential, often underpaid, usually overworked, and under-recognized professionals standing between a shovel or a backhoe and disaster. The industry faces the challenges brought on by the dilemma of paying more in the short run for a high-quality locate or cutting short-term costs while increasing the risk of damages and the associated danger. The age old adage of *pay me now or pay me later* is spot on for the locate industry. This is not to point fingers, it is merely demonstrating a reality of the locate world. The industry knows what must be done to improve things, which we will talk about in later chapters.

A locator can learn basic equipment use in a few hours, but it takes six months to a year in the field to become productive and truly understand complex challenges presented trying to locate the myriad of different types of cable and pipe in the ground.

> And I can tell you this, the locator who has spent years in the field is more than a technician. They're the living memory of the underground. They've developed instincts you can't teach in a classroom, the ability to "see" what's beneath the soil without ever breaking ground.
>
> (Jason L. Smith, LineQuest, LLC)[3]

Some of the pipe and cable may be over 50 years old and in crowded urban rights-of-ways, making tracing a specific signal very challenging. Their work may be invisible to the public, but the safety of our communities and the integrity of our infrastructure depends on them. The more training and experience they gain, the more resilient our entire infrastructure landscape will be.

> Whether you're locating one line or four, training and support are what make a quality locate.
>
> (*Tracy Purcell, Bloodhound* Top Issues Affecting Utility Locating Town Hall)

This chapter is built on research and insights from industry veterans. These experts provide a candid and thorough look into the locating profession, its challenges, its evolution, and its crucial role in public safety. Ron Peterson, President of Ron Peterson Consulting, LLC, has been in the industry for over 30 years and has had first-hand experience with every facet of locating, including doing the actual locates, training locators, owning contract locating companies, owning a utility contracting business, providing expert witness testimony relating to locating, and also serving as Executive Director of the locating industry trade association (National Utility Locating Contractors Association "Nulca").

Locating prevents catastrophic damage to underground infrastructure. Peterson points out, "We're out here protecting multi-billion dollar facilities that can kill people," Peterson states, underscoring the life-or-death stakes involved. Utility damage can cause service outages, environmental disasters, financial loss, or even loss of life.

Yet, Peterson laments the industry's lack of consistent standards and fair compensation. Not only is there a lack of a unified standard or certification, but unlike many less consequential industries that take a specific degree or government approval to enter, virtually anyone can buy locating equipment and represent themselves as a locator the next day. Not only is this a low barrier to entry, it is a risk to quality and safety, but there are few mechanisms to attract high quality and dedicated newcomers. Locator pay is typically very low, and Peterson points out that in many cases "you can make more money at McDonald's. We've got to fix that." This is a contributing factor to a high turnover rate for locators, which in turn exacerbates training issues. Peterson states that the industry is working to solve this problem knowing that, "You kill turnover by paying people what they want, giving them a career path instead of a job."

As Jason Smith confirms, "But here's the reality: we are losing too many of them. Burnout, low recognition, and lack of career development drive experienced techs away. Every time that happens, the industry loses a library of knowledge, and damages become more likely."

There are no mandatory certifications or training required for locators, but the National Utility Locating Contractors Association (Nulca), a national organization representing utility locating professionals, published its first Competency Standard for training utility locators in 1996.[4] This guideline has become an industry guide and is now in its fifth revision.[5] In 2016, Nulca rolled out its accreditation/certification program. This program is voluntary, but it does create a standard for training. There are other training certification programs in Canada, Australia, New Zealand, the United Kingdom, as well as Staking University (hereafter "StakingU") in the United States, but the Nulca program has the momentum of the industry's largest locator organization behind it. More details on these programs are in Chapter 11 and the chapter appendix.

This is not meant to imply that locators are not trained, because most companies provide in-house training or hire professional trainers to get locators up to speed. The primary issue is the lack of uniform standards for the initial training or any continuing education.

The good news is that the damage prevention industry understands this and is working toward solutions. The industry is packed full of people committed to safety and saving lives, in fact that is one of the aspects that draws people in. Creating national standards and certifications is unavoidably slow when you have to work within 50 different state laws, federal laws, regulations from the Occupational Safety and Health Administration (OSHA) and other agencies, and economic factors. Nevertheless, progress is being made!

Locate Technician Safety

Buried utilities are everywhere, and state laws create urgency, so the job of a locate technician can present safety challenges. Locating also takes concentration. Technicians in the field must be focused on their surroundings as well as their equipment, deftly handling safety risks to themselves while fulfilling their duty to ultimately reduce safety risks for fellow stakeholders further into the process. Among the many distractions and common nuisances that can range from minor to fatal include:

- Road-side and work-zone traffic
- Threats to eyes
- Confined spaces and poor soil stability
- Heat, cold, and severe weather
- Slips, trips, and falls on uneven ground
- Animal/Insect encounters (dogs, insects, snakes)
- Poison Ivy and other skin irritants and threats
- Mental fatigue and workload stress
- Dehydration, sunburn, soreness

These and other pressures help explain why this is an underappreciated job. Walking long stretches alone outdoors in the elements to mark out buried lines that someone may still strike – or that no one will dig near because the requested locating area was too large – these may deter new locators or push current locating professionals to look for another job opportunity. And while these are common safety concerns among all locators, the specific type of employment arrangement may have an impact on the workload, quality of life, and other details for locators.

One Size Does Not Fit All – Different Types of Locate Companies

Locators can be divided into three very broad categories: (1) in-house company locators who work directly for a utility company or infrastructure operator, (2) third-party contract locators commonly called "contract locators" who make contracts with individual utility owner/operators and visit sites to locate and mark their infrastructure, and (3) Private locators who are contractors hired by property owners to mark private buried infrastructure typically not owned by a public utility, like lighting cables, irrigation pipelines, behind the meter lines, and more. While there are differences in contracts, liability, and company policy, the actual work done by locators is very much the same regardless of whether they are in-house or contracted.

When people in the industry refer to locators, they are generally talking about the technicians who determine the approximate location of buried cables and pipelines with temporary marking paint and flags just prior to excavation. On large construction projects, the exact location – including depth – may be required before the project designs are done. This process is called SUE and requires an engineer to ensure detailed specifications are met. In many countries, SUE is also known as Subsurface Utility Investigation (SUI). Subsurface Utility Mapping (SUM) involves creating a map that shows where underground utilities (pipes, cables, conduits, etc.) are located using utility records along with surface locating tools and, but without necessarily doing deep engineering analysis. They are closely related but not exactly the same. Each term refers to a different scope or methodology within the broader practice of identifying and managing underground utilities.

More details on SUE in the United States will be covered in Chapter 5.

In-House Locators

Before there were contract locators, companies had to train their own staff to do locates. While this was the original model, it continues today for many utility companies across the country. This may mean that they have specific technicians doing locates full time, or they may have many technicians trained to locate when needed. The advantages of this system are that the locators only need to become experts on locating the types of cable or pipe their company owns, and typically just the type in their geographic area of responsibility. The disadvantage for those doing locates less than full time is that it is much more difficult to become an expert doing something part time.

A few of the primary reasons facility owners choose the in-house staff option are:

- The demand for locates is high enough to keep in-house technicians busy full-time.
- They have extremely sensitive facilities like high-pressure pipelines, pipelines with hazardous liquids, or high-capacity communications/data lines.
- They have complete control over the initial and ongoing training for their locators.
- They are knowledgeable of any requirements or legislation governing excavation and construction near the buried assets.

Contract Locators

Contract Locating was first started in Indiana by SM&P Conduit Company and in Arizona by STS Locating in 1979.[6,7] The premise behind contract locating was based on the logic that a good portion of the labor costs to do a locate was in getting a person and equipment to the site in a timely fashion. If the proposed construction site had buried electric, telephone, cable television, natural gas, and water pipelines in the area, all five of those utilities would have to send locators to the site. If one contract locator could get even two or three of the utilities to hire them, the efficiency gain would be significant.

Facility owners who choose to use contract locators do so for a variety of potential reasons:

- Uneven demand for locating can be hard to staff for, so using a contractor eliminates that problem for the facility owner.
- They may not have enough demand for full-time staff to do locates. They may feel a contract locator who has full-time locators can do a better job versus their own staff who only locate sporadically.
- Their buried assets may be spread out geographically, making it hard to cover with internal staff.
- They believe a contract locator may offer better training, because locating is the main focus.
- They may feel that hiring a contract locator will reduce the risk of liability. Ron Peterson, who has been an expert witness in hundreds of cases, says "In most cases, [shifting liability to a contract locator] probably doesn't work."
- Consortiums of distribution utility companies initiate locating contracts that dispatch a single locator to locate and mark all distribution assets within the digsite rather than each utility company dispatching their own to achieve an economy of scale and process.

Private Locators

Utilities that operate in public rights-of-way locate their buried facilities upon request. They typically mark up to the service meter. These requests often come through the local 811 Center, if they are members. However, a large percentage of buried cables and other pipelines in the United States are privately owned and may not be part of the 811 Center System. State laws vary, but some buried infrastructure systems are not required to be in the state's 811 Center program either based on formal exemptions or the small footprint of their facilities. There are few to no official sources with definitive data on the total number of miles of buried facilities in the United States or an exact breakdown on what is covered by 811 Centers. However, the most common estimate of the breakdown between utilities covered by the 811 system is around 35 percent, with the remaining 65 percent being private and other facilities.[8] Complicating this further is that the estimate is for entities, not their facility mileage.

Many organizations in the industry use a figure of "over 20 million miles" of buried pipes, cables, and wires, which is also over 100 billion feet, an impressive figure to help average Americans conceptualize the enormity. But what exactly is included in this estimate – which we explore more in Chapter 4 – is not clear. The challenge is that much of what is below the ground is what are considered public utilities, yet many more lines are on private or restricted areas or are behind-the-meter lines. Not every locator is equipped to identify every type of facility, and standard 811 tickets only cover a certain portion of all buried infrastructure.

Private locators are typically hired by the property owner to locate buried facilities that are not owned by a member of the local 811 system. Private locate companies normally have more equipment and tools to locate difficult-to-find facilities. They are likely to have both electromagnetic and ground penetrating radar (GPR) tools, as well as using sondes and vacuum excavation. Because they are hired by a property owner, they are not usually pressured by the time constraints of an 811 locate request.

These are some of the reasons that companies hire a private locate company:

- They know there are buried facilities on their property that are not being located through the 811 system, and they want to avoid damaging them during an excavation project.
- They are creating detailed plans based on SUE in order to design a major project.
- They are creating a map of all the buried facilities on their property.

Some examples of facilities that are likely to use private locators are refineries, hospitals, campuses, military bases, and airports.

Two personal experiences that should have required private locators help demonstrate the settings and scale of impact when communication or coordination breaks down beneath the surface. Each of the following anecdotes come from Scott's decades-long career:

We were holding a damage prevention conference and expo with digging demos at the Los Angeles County Fairgrounds, so we called in a locate request to DigAlert (the 811 center for Southern California). We also asked the Fairgrounds for details on any buried facilities they had where we would be digging. The Fairgrounds Operations Manager said there were no cables or pipelines, and he said they did not have a map.

We were going to be digging in the center of the horse racing track which was surrounded by lights. This did not seem right to me. Luckily, my uncle used to be the Operations Director at the Fairgrounds, so I called him. He said that there were tons of cables and pipes under the racetrack infield, including a high-voltage cable and fiber optic cable. I went back to the Operations Director and asked him to sign a release assuming 100 percent of all liability for any damage or injuries that might happen. Magically, at that point, he found a map and we were able to avoid disaster.

In a similar situation, we were holding another damage prevention conference with digging demos at the Rio Casino in Las Vegas. They gave us permission to dig up the asphalt parking lot to do a trench shoring demonstration. We put in a ticket with USA North 811 and met with the Rio Operations Manager. The Operations Manager said there was nothing under the parking lot where we were digging, so we did not hire a private locator. During the excavation, the contractor saw some buried marking tape and stopped the backhoe operator. It turns out there was a high-voltage cable there, and below that was the fiber optic cable that connected the sports betting for all the Harrah's casinos! This took place during the NCAA Basketball's Sweet 16, which is one of the busiest times of the year for the sports betting business. One more shovel full from the backhoe, and he would have cut them both. Thank God the contractor was well trained on safety and damage prevention, and that someone decided to spend the extra pennies to bury warning tape above the cable.

The North American Private Utility Association (NAPUA) fills the structural gap illustrated by incidents like those above. While 811 systems govern publicly owned utility infrastructure within the right-of-way, they do not extend to the vast network of privately owned buried utility infrastructure located on private property. Grant Piraine, Founder of NAPUA, states that after more than three decades working in ground disturbance, private locating, utility coordination, and risk management, he has consistently observed that this gap results in inconsistent practices, unclear responsibility, limited records, and elevated safety and liability exposure.

NAPUA advances its mission through a Life Cycle Model that recognizes three sequential lanes of risk management: SUE during the design phase, public utility governance within regulated 811 systems, and the use of private locators for private property locates and public verification locates before ground disturbance execution. The third lane, where excavation occurs on private property, currently lacks a coordinated North American guidance framework and represents the final operational checkpoint before physical impact. By focusing on this execution phase, NAPUA's Best Practices and companion Guidelines complement existing public and engineering systems and strengthen overall industry safety and damage prevention practices.

The Evolution of Locating Tools and Technologies

At some point in time, you have probably noticed a person wearing a bright vest walking along with what might look like a metal detector looking for treasure on city streets. At the same time, they are spraying colorful paint on the ground or putting down matching colored flags. If you have not noticed this before, we guarantee you will see them now, because once the idea is in your head, you will start seeing them everywhere. This kind of spray paint is ubiquitous.

In most cases, these "wands" are sophisticated electromagnetic instruments that replaced forked sticks or L-shaped "witching rods." Yes, these really did exist as a way to find gas and water pipelines even up until the 1960s, and some say they still do in some isolated places.

Locating isn't just a modern trade; it's part of a centuries-long pursuit to bring the invisible into view. The evolution of the tools tells its own story:

1500s–1800s – Witching Sticks/Rods: Used by miners, well-drillers to locate veins, springs, later iron water-mains because no other geophysical tools existed. There are reportedly still isolated cases of people using this method, but I wouldn't want to go to court citing this as the method I used to locate a natural gas pipeline.[9]

1831 – Electromagnetic Induction: Michael Faraday, an English chemist and physicist, showed that a changing magnetic field can generate electricity. In experiments with coils of wire on an iron ring, he proved that switching current in one coil induced current in another. He submitted his findings to the Royal Society in London, establishing the principle of electromagnetic induction.[10]

1910 (cerca) – Faraday's induction: German telephone engineers photographically documented the use of a hand-wound induction coil on a wooden truss to trace buried power/telecom cables. This was the first recorded field-application of Faraday's induction principle to underground utilities.[11]

1920–1931 – Metalloscope: Dr. Gerhard R. Fisher adapted an aircraft radio direction-finder into the handheld "Metalloscope." This was the first portable device that a single operator could carry; marketed from 1931 and patented in 1933 (US 2,066,561) specifically to find "buried ore, pipes, or the like."[12]

1959 – Vacuum excavation: This is the process of loosening the soil with an air lance or low-pressure water, then removing it with a big vacuum hose. Picture an industrial strength shop-vac that leaves a neat, narrow "pothole" that's only as wide as a coffee can. This is the only way to determine the exact location of a pipeline or cable. This is often called "daylighting" or "potholing." The first vacuum system for locating was launched in 1959, when Underground Services (now SoftDig) introduced its air-vac truck specifically for non-destructive vacuum excavation with a focus on attaching anodes to gas pipelines and locating utilities. Commercial sales exploded in the early 1990s.[13]

1964 – Twin-aerial Receiver: Bell Labs developed the "Depthometer," a twin-aerial receiver that gave both horizontal location and depth. This became the Bell System (two decades before the breakup of AT&T) standard across the U.S. telecom network and set the accuracy template for modern locators.[14]

1968 – Ultra-high-frequency locator: Pipehorn patented the first ultra-high-frequency, stand-up locator. This solved the problem of cast-iron gas mains and proved that rugged, one-person tools could live in a service truck.[15]

1972 – Magnetometers: The true commercial birth of magnetometer-based utility locating can be traced back to the release of Schonstedt's hand-held magnetic locator. This meant that technicians could now throw a 2-lb tool in the truck instead of hauling geophysical gear. Magnetometers are fast, passive, and require no connection point. They can be used to locate cast-iron and steel gas/water lines laid before tracer wire existed, as well as valve lids, curb stops, and manhole lids that have been buried or paved over.[16]

Late 1970s – Electrolocation: Radiodetection brings the twin-aerial concept to mass production and began exporting RD-series locators worldwide. This marks the start of truly commercial, off-the-shelf locators for municipalities, power utilities, and contract locating firms.[17]

Late 1970s – Sonde Transmitters: Sondes (mini-transmitters) have been around since the early post-war years and became common with sewer-TV rigs in the late 1970s.[18]

1970s–1980s – Ground Penetrating Radar (GPR): Geophysical Survey Systems Inc. (GSSI) built the first commercial ground-penetrating-radar unit. As GPR units got smaller, lighter, and less expensive in the 1980s, GPR was adopted for use as a utility locating and mapping tool. By the early 2000s, GPR became a standard tool for more and more locators.[19]

Early 1980s – Dr. Earl Peterman's use of multiple peak antennas: This provided depth readings on the Metrotech 810, which was the first commercial use of depth on locating devices.[20]

Late 1980s – EM Standard: By the end of the 1980s, EM induction had become the de-facto global standard for metallic utility detection.

Early 1990s – Acoustic excitation ("knocker"/"pulser"): A small piston or valve is clamped on the line (or hydrant), which rhythmically bangs or pulses the water/gas column to track the sound with a ground microphone or leak logger. A ground microphone hears the peak sound and you follow the noise.[21]

Early 1990s – Pulsed-water EM conduction: A "pulsed-water transmitter" screws to a tap or hydrant, injects a modulated current into the conductive water column so a standard EM locator can follow the pipe. The locator electrically energizes the pressurized water column, so it behaves like a long, wet conductor. A standard EM pipe-and-cable locator then follows the electromagnetic field.[22]

1990s – Thermal or IR imaging: Thermal imaging joined the utility-locator's kit in the mid-1970s via aerial surveys that spotted hot district-heating mains, but it didn't become commonplace until small, affordable IR cameras arrived in the 1990s. The camera simply looks for tiny surface temperature contrasts caused by the pipe or its leaking product.

1990s–2020s – Multi-frequency, current-direction, GPS-integrated locators: These instruments offered more detailed information, but the underlying Faraday/Bell-Labs method remained unchanged.[23]

2001: CCTV lateral launch from main line cameras were developed and used for locating gas pipelines and cross-bores in sewers.[24]

Mid-2000s – Duct rodders with copper core: A copper conductor is pultruded inside a fiberglass rod allowing any EM transmitter to energize the rod itself, turning the whole length into a traceable path. This technology was pioneered by Jameson.[25]

2012 – Dielectric/UHRF scanners (e.g., AML Pro): Think of a dielectric/UHRF locator (like the AML Pro) as a tiny, ground-facing "flash-light" that shines very-high-frequency radio waves (mini-microwaves) straight down into the soil and then listens for how much of that light splashes back. It operates much like a stud finder one might use at home. This technology measures the dielectric contrast between soil and plastic pipe, conduit, or clay tile. This technology works in wet clay where GPR does not work. The first hand-held dielectric/UHRF utility scanner reached the market in 2012 with SubSurface Instruments' AML.[26]

Today two technologies dominate the industry: Electromagnetic (EM) and Ground Penetrating Radar (GPR).

Electromagnetic (EM)

EM pipeline and cable locators are essential tools for locating underground utilities like pipelines, power lines, and communication cables.[27] They operate by identifying electromagnetic fields associated with these utilities. In most cases, this involves sending a radiofrequency signal down the pipeline or cable. If the cable or pipeline is not metal, there needs to be a metal wire, known as tracer wire, buried above or beside the pipeline or cable in order for an EM locator to find it. Some communications cables will have a metal sheath to protect the cable or a metal strength member which can also transmit the radiofrequency. Details on tracer wire are covered after the EM locator section.

While the concept behind using an EM locator may seem simple, it is far from easy. Most locating experts, like Ron Peterson, believe it really takes a year or more for a locate technician to become a top notch locator. A few of the technical issues a locator faces and needs to learn about include:

- Signals jumping from one cable or pipe to another in crowded rights-of-way.
- Different soil conditions.
- Non-metallic pipelines and cables without tracer wire.
- Broken tracer wire.
- Distortion and its effects on locate accuracy.
- Finding the correct signal when others exist on the site.

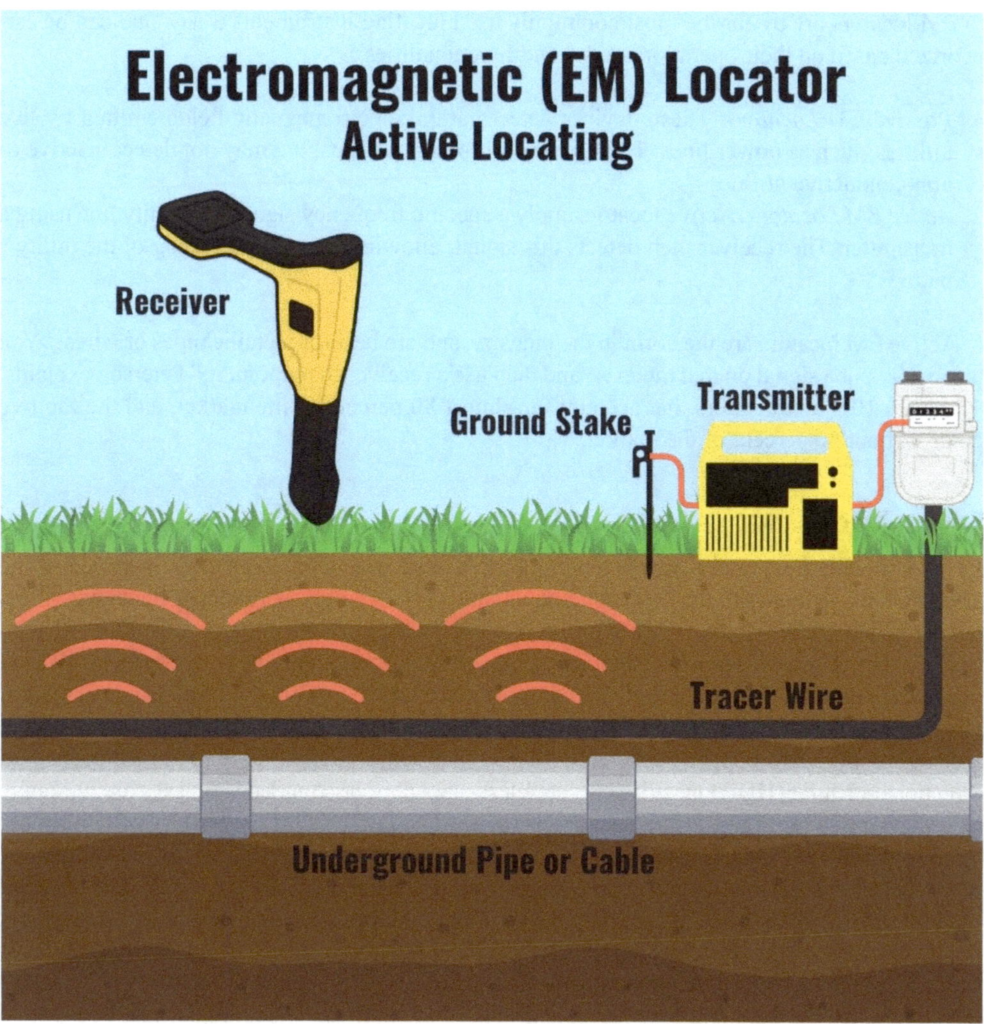

Active locating electromagnetic (EM) locator.

Coauthor Scott has sat in on a number of EM locator training workshops. While the principles may be simple, truly grasping it in the real world is not simple. There are rarely two situations that are exactly the same. Based on being around locating and locators for over 30 years, I firmly believe locating is a mixture of science, logic, and detective work; and in the end, it really becomes an art.

I can say with 100 percent confidence that truly learning to locate is not done just in a classroom or by watching videos. It needs to be learned in the field by being exposed to hundreds of different situations. On top of that, all this may have to be done in the rain, blazing heat, freezing cold, and possibly out on a busy road. Never look at paint, flags, or "whiskers" again without acknowledging the skill and dedication it took to get the locate done in order to help keep excavators safe.

EM locators are by far the most commonly used locating instruments today, and can be categorized based on their operation modes and functionalities:

1 *Passive EM Locators*: These devices detect natural electromagnetic fields emitted by live utilities, such as power lines. They are useful for quick scans, but may not detect inactive or non-conductive utilities.[28]
2 *Active EM Locators*: Active locators apply a specific frequency signal to a utility line using a transmitter. The receiver then detects this signal, allowing for precise tracing of the utility's path.[29]

Active EM locators are the norm in the industry, and are best for metallic pipes or wires. "You physically put a signal on that metal ... and then use a receiver to pinpoint it," Peterson explains. There are 10 manufacturers that account for almost 80 percent of the market, and the top five control about 50 percent of the market.[30]

Tracer Wire

In the United States, tracer wire is simply an insulated wire (conductor) that is buried in the same trench as a non-metallic pipe, conduit, or cable. Interestingly, in Australia tracer wire is often not insulated. When the utility needs to be found, a locator transmitter is clipped to the wire and to a remote ground stake; the alternating current that flows along the wire sets up an electromagnetic field that an EM locator can trace from the surface exactly the same way technicians trace a metallic pipe or power cable.[31] There are many types of tracer wire, but the most common are solid copper and copper-clad steel (CCS). When extra strength is required for horizontal directional drilling (HDD) installations, stainless-steel core or stranded stainless core wire may be used to avoid breakage during installation. More detail on HDD is covered later in this book.

For tracer wire to be effective, it must be installed two to six inches above the pipe or cable.[32] Since the locator is actually locating the tracer wire, not the pipe or cable, it needs to be installed in close proximity to the utility. This again shows the importance of tolerance zones. It should not be wrapped around the cable or pipe, because it will cause a distorted signal. Many specifications call for the wire to avoid contact with a gas pipeline to reduce the chances of lightning traveling down the wire and causing an ignition. The wire must be continuous and be repaired if it is broken or it will not work.

Today, tracer wire is installed with almost all non-metallic pipelines and cables, but it has not always been the standard. Tracer wire became a federal requirement for new plastic natural gas mains and service lines in 1996.[33] Beginning in 2014, some states began requiring that all underground facilities be "electronically locatable," and other states required tracer wire for non-metallic water and/or sewer lines.[34]

Chris Ross, Asset Protection & Relocation – Principal at Telstra in Australia, says that tracer wire is not required in Australia, so "we've been investigating IMU/Gyroscope-based pipe mapping technologies for locating empty conduits or those with just optical fiber." Here's an example.[35]

Many old sewer lateral pipelines that go from the house to the mainline were non-metallic, and tracer wire was not used, making them difficult to locate. This led to many accidental cross-bores. The official definition of a cross-bore is "the intersection of an existing underground utility or structure by a second utility installed (typically with trenchless technology), resulting in direct contact that compromises the integrity of one or both utilities."[36] The most serious issue

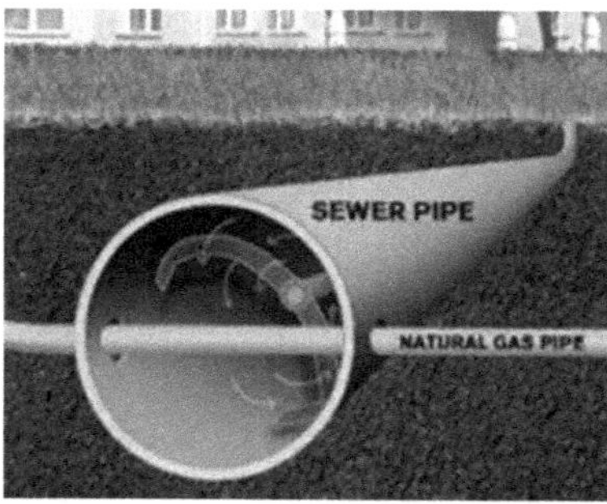

A cross-section view of sewer lines being intersected by natural gas pipelines.

caused by cross-bores is when a natural gas line is inadvertently installed through an unmarked sewer lateral. This can lead to a backup of the sewer lateral, which can lead to a service running an auger through the line to remove the blockage, severing the gas line. The gas can then leak into a building via the sewer, later causing an explosion if a pilot light or an electrical switch is engaged.

The cross-bore problem was first officially identified in 1976 in a DOT report.[37] The natural gas industry began developing standard practices to reduce cross-bore incidents, eliminate existing cross-bores, and educate plumbers of cross-bore risk when using sewer cleaning machines. Now, there are standards, Leading Practices. and guidelines designed to find and eliminate cross-bores. Though gas pipelines in sewers have been catastrophic, intersections with electric lines have also been deadly, and other utility cross-bores can result in costly damages as well.

"Call Before You Clear" (CBYC) was created to address the issue of legacy cross-bores.[38] A voluntary, multi-utility program that asks plumbers, drain-cleaners, and homeowners to call the local natural-gas utility (or 811) before mechanically clearing a clogged sewer/septic line, so the utility can run a quick records check or free camera inspection and rule-out a gas-sewer cross-bore. "Call Before You Clear" is the post-2010 safety counterpart to "Call Before You Dig."

It began as utility-specific outreach (Washington Gas 2008), expanded after the St Paul 2010 explosion into a multi-utility Minnesota campaign, and by 2014 had become a

CCTV captures of natural gas pipelines cross-boring through sewer lines.

CGA-endorsed best practice now embedded in 811 workflows from Pennsylvania to Arizona.[39] The program does not replace cross-bore-prevention rules, but it fills a critical gap: catching hidden gas-in-sewer conflicts before a plumber's cutter turns them into the next headline explosion.

Seeing the process and consequences for yourself may be the most impactful way for this to stick:

Call Before You Clear

Ground Penetrating Radar (GPR)

GPR works like underground sonar using radio waves.[40] GPR equipment is used in a wide variety of industries, but in the damage prevention industry, its most common use is locating non-metallic pipelines without tracer wire and cable without a metal sheath. As the locator pushes a cart (like a lawn mower) across the ground, a radar antenna sends waves into the ground. When those waves hit something different, like a pipe or any subsurface structure, they bounce back. A computer on the GPR cart shows that bounce as a shape or line on a screen. A trained technician can look at these images and be fairly certain if it is showing a pipeline, cable, duct, or anything else that should be marked as a part of the location.

GPR equipment combines both high-frequency antennas, which show small, shallow objects like cables near the surface and low-frequency antennas that look deeper into the ground, but they provide less detail. The design intent – combining both types of antennas on one cart – is to collect data from both antennas simultaneously, enhancing efficiency and providing a more complete subsurface profile without the need for multiple surveys. In some cases, single-frequency systems may be more appropriate, especially if the survey focuses exclusively on either very shallow or very deep targets, but that is not common when locating buried pipelines and cables.

The fact that GPR sends a short radio-wave pulse into the ground and "listens" for echoes means the soil conditions impact how far that pulse travels and how clear the echo is.[41] The soil's condition, also referred to as the dielectric constant, tells us how easily an electrical signal or radar wave can move through the ground.

These are some situations where GPR works well:

- Dry, sandy, well-drained soils like deserts (Arizona, New Mexico, etc.), sandy areas, and power line corridors.
- Loose gravel or crushed stone backfill, like recently built subdivisions, utility corridors, railroad rights-of-way.
- Dry or moderately dry silts/loams.
- Pavement

Expert locators can often "read" the site and know which tool they'll need. They also have intuition and knowledge built up over years to help look for additional context clues and site characteristics that aid in their work.

> While Utility locate practitioners may not see the pipes, we often see the trench and this helps pin-point locations for further subsurface investigations.
>
> (Michael Twohig, Director of Subsurface Utility Mapping at DGT Associates | Utility Mapping, Surveying)

GPR does not work as well in these conditions:

- Wet clay and silty clay loams, and soil with metallic content
- Salt-contaminated or tidal soils like coastal marshes.
- Very wet, fine sand
- Rebar-dense or fiber-reinforced concrete

Overview

- Dry, sandy, gravelly: thumbs up, expect good depth and crisp pipe images.
- Wet, clay-rich, salty: thumbs down because GPR may only "see" the first foot.

> GPR has the potential to identify things that other tools just can't – concrete pipes, plastic, abandoned trenches. That makes it vital to our industry.
>
> (Dan Bigman, Bigman Geophysical)

Price Range

Generally, quality EM locators used in 811-related contract locating range from about $1,000 to nearly $10,000, with many professional units falling in the $3,000 to $8,000 range and some specialty configurations costing more.

Professional GPR locators generally start in the low-to-mid tens of thousands of dollars and extend above $30,000, with many common utility-locating systems falling around $15,000 to $25,000.

Magnetic locators commonly used in utility locating come in as low as the hundreds and often range up to about $1,500 to $2,000, while more specialized magnetometer systems may cost substantially more.

Vacuum Excavation Is a Vital Part of Locating

Vacuum excavation can be used for many things, including rescuing someone buried due to a trench collapse, but in this chapter, we are focusing on using vacuum excavation for verifying the location of buried cables and pipelines.[42] This process is called "potholing" or "daylighting." This process creates a round hole, six inches to a foot in diameter, in the ground all the way down to the location of the pipeline or cable. Daylighting comes from the fact that the hole puts daylight on the pipeline or cable. Per James Anspach, one of the pioneers of vacuum excavation, the name it actually morphed from daylighting to potholing when users did not compact the repair, leading to a "pothole" type damage.

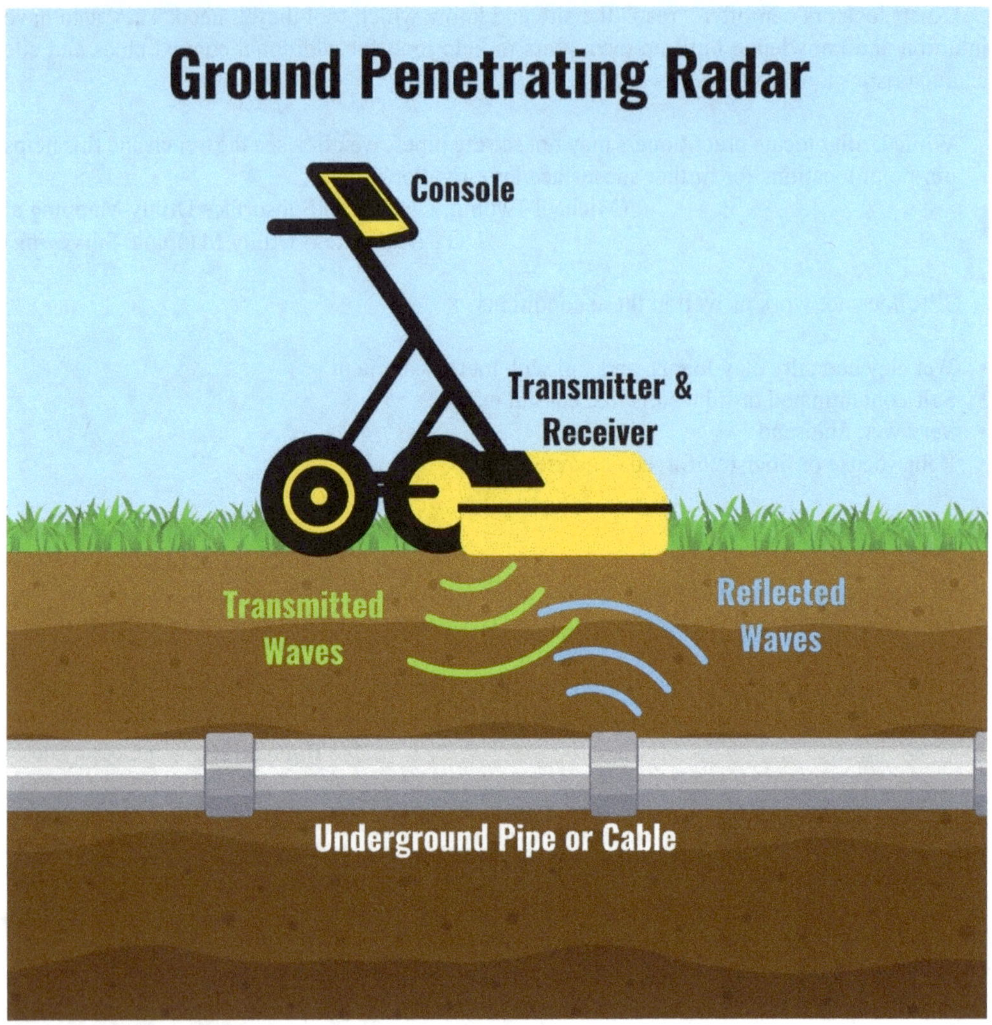

Ground penetrating radar (GPR) system.

Vacuum excavation can be done using either water or air to create the hole.

The water-based system is called hydrovac, and it uses a pencil-thin jet of high-pressure water, very much like the pressure washer you use to clean a driveway or wash your house. To loosen the soil, the jet turns the ground into a muddy slurry that is sucked up the hose into a holding tank on the truck. Digging with air uses compressed air pushed through an "air lance" to fracture the dirt. Just like with the hydrovac process, the dirt is vacuumed straight into the tank.

Hydrovacs create slurry, which must be disposed of per the local regulations that often include licensed hydrovac-waste facilities, municipal dewatering pits, or an approved landfill that accepts "special waste." Air-vac spoil is normally just dumped out of the tank and put back into the pothole, then tamped. If the hole was through asphalt or concrete, the surface is repaired to match what was there.

EM locators provide the estimated location of the buried facilities, and the paint and flags provide a visual path for the cable or pipeline. But putting eyeballs on the pipeline or cable itself by exposing the facility is the only way to know the exact location. Potholing can be the fastest, most economical way to do this, depending on the depth, soil conditions, and type of facility. This is one reason why there are tolerance zones where only hand digging or vacuum excavation is allowed.

There are many circumstances where potholing or daylighting is critical and often mandatory.

- High-consequence lines like high-pressure gas, hazardous liquid, 345 kV feeders, fiber backbones. Pipeline and electric-transmission owner standards typically demand daylighting 15 feet on either side of the crossing and at each tie-in.
- Where there are un-locatable plastic pipelines
- Abandoned lines
- Trenchless crossing (HDD, auger-bore, pipe-burst) of existing plant
- SUE – Design or survey needs “Quality Level A” data (details are covered in Chapter 5)
- If you need exact depth

Potholing can also be the best form of excavation when only a small area needs to be excavated, like:

- Fence post installation
- Utility pole installation
- Sign post installation

How Much Does Potholing Cost?

The costs vary widely, but commonly cited industry estimates range from $600 to $900 per hole. The cost per hole can get down to near $300 or above $2,000. The factors that impact the costs – both per hole and a project total – help explain why the range is so great:

- The largest factor is when data on the exposed utility is required.
- How many holes are being excavated.
- The depth and size of each hole.
- The proximity to water (when using hydro excavation) and disposal site.
- Disposal costs (for the soil extracted from the site).
- Traffic control costs if in the roadway.
- The soil type impacts the price because it impacts the speed of the work.
- The surface material (dirt, concrete, asphalt, etc.).
- Reinstatement costs.

When potholing is not required, the choice becomes a risk-based decision weighing the known cost of potholing versus the potential cost of a damage. The types of facilities (high-voltage cable, fiber optic cable, telephone drop for a home, etc.) in the dig area impact the decision, as well as the facilities they are serving (hospital, major business, etc.). Communication between the excavator and facility owners is the key to making the best decision.

Michael Twohig adds:

> In my experience, the project owner and their representative undertake a project Risk assessment for groundbreaking. In many cases, a risk-transfer mechanism is used to trust the contractor to perform "test pits" in the early stages of the project. A proactive team utilizes vacuum excavation to determine the precise horizontal and vertical locations of critical networks, enabling the refinement of final civil design decisions.

Emergency Locates

In most states, everyone is required to request a locate two to three days before excavating. However, there are emergency situations for which the locate needs to be done much more quickly to protect property and life. In Illinois, the legal definition of an emergency locate is when there is "imminent danger to life, health, property, or a utility-service outage that requires action before the normal 2-day waiting period expires."[43] A broken gas main, a severed power line that feeds a hospital, or a water-line break flooding a street all qualify. Digging a fence-post the same afternoon because the contractor showed up early, or last minute tent installation because of rain do not qualify.

Scott had to have a five-foot section of sewer lateral replaced because a tree root had grown through it. The contractor showed up without any paint or flags on the ground, so I asked him about it, and he said, "I call the locators, and they come right out to mark it." I explained that no, it doesn't work like that. He said that because it was to repair a sewer lateral, it was an emergency, but I told him that was true for a sewer lateral that was leaking, not one that was simply being proactively replaced. He was shocked, but happy when I told him all he had to do was put in the request 48 hours (excluding weekends and holidays) in advance for replacement jobs. Damage prevention is almost always about communication.

When a ticket is deemed to be an emergency locate, it must be completed in two to three hours in most states, but the laws vary by state. The more these are used, the more they can strain the industry and force difficult prioritization of locate jobs. Because of this, some states allow the facility operator to charge the excavator for false emergencies. Some of this requires additional personnel, even while other types of scheduling and technology can help reduce strain. Getting those bodies is not always easy.

> Competing initiatives are drawing resources in a dozen directions ... at the end of the day we have a manpower issue.
>
> (*Eric McLeish, Director Operations, NPL Construction* Late Locates Town Hall)

Some of those competing initiatives are policy-generated, while others are simply based on the competitive market.

Market Forces and Liability Shifts

One reason it is difficult for contract locating firms to pay higher wages to locate technicians is the competitive bidding process when trying to win large contracts. Many facility owners do not have a good way to factor in the quality of the locates into the bidding process, which drives the cost per locate down. The direct overhead for each locate is made up of labor cost for the technician, paint and flags, and transportation costs to get to and from the locate site. There are plenty of fixed costs including vehicles, locating instruments, and general overhead, but when looking at the direct per-locate ticket costs, the labor cost is by far the largest single cost.

This cost structure without any way to tie the cost of damages due to an inaccurate locate, makes it more difficult to increase locator pay and remain competitively priced. There is no specific entity to point the finger at, but it is a challenge for the industry. When Ron Peterson owned a contractor locating company, he described an example of this: "We lost a contract to a cheaper competitor. They had more damages in 30 days than we had all year. We were brought back in, but the customer wanted us to lower our price."

Sometimes the person making the decision on which locator to use is only evaluating the cost of the locate, because they believe they can push back any damage costs onto the excavator or locator. Peterson also exposes a legal myth that liability can be transferred to the contractors. "It's a liability push. When it comes to big incidents, it doesn't work. Utilities try to push it back, but in gas explosions and deaths, that doesn't hold up."

The Cost of a Locate

The cost of a locate generated through an 811 Center is different from a locate being done as part of a SUE project or a private locate on a military base, university campus, etc. Private locates are often requested directly from a project planner to the locating company and are generally billed at an hourly rate for a specific locate need. General excavator locate requests for marking public utilities go to the 811 Center, and these costs are typically priced *per ticket*.

With a One-Call locate, the excavator requests a locate through the 811 Center. The Center creates a locate "ticket" with a unique ticket number assigned to it for that request. The 811 Center then notifies its affected members by transmitting the ticket digitally. All members who receive a transmission regarding the project receive the same ticket.

Because locate requests go to 811 Centers, there is often a misconception that these Centers actually do the locates. *811 Centers DO NOT DO LOCATES; they receive locate requests, create tickets, and forward those tickets to their members. The member utility companies either do the locate (in-house) or hire a locator (contract locator).*

There are several costs to separate in this process. The excavator making a locate request pays nothing. The 811 Center charges each utility company receiving a ticket a nominal fee, which may be around a dollar but varies. These are known as "outgoing transmissions." The DIRT report explains that each inbound request generates four to six outgoing transmissions to individual facility owners which are members of the 811 Center. Picture a single residential lot, with one locate request but five utility lines (e.g., water, gas, electric, fiber, and sewer), so the activity handled in the field is far larger than the ticket count.

Once the utility company is notified that someone is planning to dig in the vicinity of their lines, and having already paid the nominal cost to receive the ticket in the first place, they have to decide how to handle the locating cost itself. The average per ticket locating costs vary significantly depending on the annual volume and geography. The per ticket price range in 2025 likely ranged from \$12 to \$65 for standard tickets; that is the amount a utility company may pay a locator company to go mark a site to fulfill a single locate request. The cost for emergency locates can be higher, as well as the cost to locate critical high-risk facilities.[44]

These costs become very significant because, according to the 2023 CGA DIRT report, there were 39.97 million incoming locate requests in the United States, plus another 2.37 million in

Canada.[45,46] With new fiber optic buildouts, constant construction, seasonal peaks, and workforce turnover, among other factors, matching market dynamics with accurate and timely locates is an expensive task. Finally, with all these costs in mind, there is the paycheck cut to the man or woman in the field making it all happen.

Locator Pay

Locate technicians are paid on an hourly basis, and current pay is an average between $20 and $30 per hour. If the contract locator company charges the utility company $12 to $65 per locate, for instance, more than half of their charge is going to labor costs, which is why the company benefits from scale by contracting with multiple utility companies to locate their lines on the same job.

The locator, however, may be expected to be out in the elements locating dozens or hundreds of buried lines each day for the same hourly rate regardless of their workload. While this pay may be competitive relative to other industries, the pressures and expectations alongside working conditions can feel limiting. The flip side of this is that the job allows for incredible independence, work in the great outdoors, and fascinating use of technology to develop mastery of a vital skillset – not to mention above all the locator is helping save lives.

Locator-Specific Challenge and Future Opportunities

The explanation of the locate industry and dynamics it faces detailed above do not yet scratch the surface of the many real-time challenges and innovations taking place. As a core pillar of the overall damage prevention industry, the locators are simultaneously depended on and unthanked for their work. Turning a famous line by Winston Churchill a bit, *never was so much owed by so many to so unknown a group.*

As this book proceeds and the virtual conference advances, locators will feature prominently. So too will additional challenges and hurdles (covered in Chapter 9) and many imminent solutions (Chapters 8, 10, 11, and 12). Many of the issues create ripples throughout other segments of the industry as well, while solutions often exist at the intersection of two or more stakeholder groups.

While real challenges remain – from the time pressures created by state laws and undefined ticket scope, to outdated records, abandoned lines, and the need for consistent practices – the resilience of the locating profession continues to shine through. Locators adapt daily to shifting workloads, evolving technology, and the complexities of underground infrastructure, and their work underscores both the difficulty and the importance of doing the job well.

Encouragingly, conversations about certification, white-lining, positive response standards, and even new tools like AI-enhanced mapping or drone-mounted GPR all reflect a forward-looking industry eager to improve. Not every idea will be universally embraced, but the willingness to innovate and refine the process is itself a mark of progress. The chapters ahead will explore these opportunities in greater depth, with the aim of showing how today's challenges can be the foundation for tomorrow's solutions in damage prevention.

Locators remain one of the most critical and ongoing *solutions* in the industry. When locators get it right, safe excavation begins in harmony. When they don't, the clock starts ticking toward chaos.

Chapter 2 Resources

Nulca is the industry association for locating professionals

Locator Safety and Appreciation Week (LSAW)

ESA/ACTS Now Town Halls on locating topics. The library expands monthly.

American Locator Magazine News and Video Archive – Planet Underground

Locator Training Resources to Search

- Nulca Accreditation and Training
- Staking University provides a wide variety of locator training programs
- Utility Training Academy provides locator training classes
- North American Private Utility Association (NAPUA)
- CERTLOC Certifications is an Australia-based program
- CAPULC Training is a Canadian-based training program
- Alternate Locate Service Provider program created by Utility Safety Partners specifically for Alberta

Notes

1 American Public Works Association (APWA). (2025). Uniform-color-code-1.pdf. https://www.apwa.org/wp-content/uploads/Uniform-Color-Code-1.pdf
2 Hoffer, T. (2017, September 22). *Understanding the Tolerance Zone and Employing Best Practices During Excavation.* SoftDig. https://www.softdig.com/blog/understanding-tolerance-zone-best-practices-excavation/
3 Smith, J. L. (2025, October). *The tenured utility locator, the backbone of damage prevention* [LinkedIn post]. LinkedIn. https://www.linkedin.com/posts/jason-l-smith-180448194_utilitylocating-callbeforehoudig-damageprevention-activity-7374410404489306112-DHcm/
4 NULCA. (2025). About NULCA. https://nulca.org/about/
5 NULCA. (2017). Professional Competence Standards for Locate Technicians. https://www.energy.nh.gov/sites/g/files/ehbemt551/files/inline-documents/enforcement-nucla-competency-standards.pdf
6 Utility Protection Trainers LLC. (2025). Utility Protection Trainers. https://utilityprotectiontrainers.com
7 Interview with Ron Peterson.
8 UtiliSource. (2024, May 30). Call before you dig: How does 811 work? https://utilisource.us/how-does-811-work/
9 See Scan. (2021). Dowsing Rods: Magic, Myth, or the Mind? https://www.seescan.com/wp-content/uploads/2021/01/Dowsing_Final.pdf
10 Britannica. (2026, February 2). Faraday's discovery of electric induction. https://www.britannica.com/science/electromagnetism/Faradays-discovery-of-electric-induction
11 Radiodetection. (2017). The theory of buried cable and pipe location. https://www.radiodetection.com/sites/default/files/Theory-cable-pipe-location.pdf
12 Fisher, G. R. (1937, January 1). Metalloscope. https://patents.google.com/patent/US2066561A/en
13 Hoffer, T. (2018, November 19). Vacuum excavation: A complete overview. https://www.softdig.com/blog/vacuum-excavation-overview/
14 Lai, W. W. L. (2019). Specifications for Nondestructive Testing, Surveying, Imaging, and Diagnosis for Underground Utilities. https://www.polyu.edu.hk/lsgi/uusspec/images/Publications/NDTSID-UU-1.1_PCL_draft_20190316.pdf
15 Pipehorn Locating Technology. (2025). About Us. https://pipehorn.com/about-us/
16 Stocking, A. W. (2005, June 30). Industry insider: Schonstedt: Finding new paths. https://amerisurv.com/2005/06/30/industry-insider-schonstedt-finding-new-paths/
17 Radiodetection. (2017). Rd7100 ® precision locators. https://www.radiodetection.com/sites/default/files/product-downloads/RD7100-brochure-V2.pdf
18 IBAK. (2022, December). IBAK Sewer and Manhole Inspection Systems. https://cnilocates.com/wp-content/uploads/2022/12/Ibak-Crawler-Products-Brochure.pdf
19 GSSI. (2020, March 4). GSSI's Innovations Through the Years – 1970–1989. https://www.geophysical.com/gssis-innovations-through-the-years-1970-1989
20 Peterman, E. J. (1983, June 7). Apparatus for determining the distance to a concealed conductive object which is radiating an alternating current signal. https://patents.google.com/patent/US4387340A/en
21 Huebler, J E, Campbell, B K, & Ching, G K (1993). Active acoustic technique accurately locates buried PE pipe. Pipe Line Industry. https://www.osti.gov/biblio/5921298
22 Radiodetection. (2004, April). RD500 OPERATION MANUAL. https://www.radiodetection.com/sites/default/files/RD500_Manual.pdf
23 Radiodetection. (2017). The theory of buried cable and pipe location. https://www.radiodetection.com/sites/default/files/Theory-cable-pipe-location.pdf
24 CUES Inc. (2019). History. https://cuesinc.com/pages/history
25 Jameson. (2021, November). DUCT HUNTERTM TRACEABLE RODDERS. https://www.lee-jensensales.com/wp-content/uploads/2021/11/Site-Pro-2021-Catalog.pdf
26 SubSurface Instruments, Inc. (2012, June 6). Subsurface Instruments, Inc. (SSI) Unveils the AML, An Underground Locating Game-Changer. https://www.prlog.org/11894069-subsurface-instruments-inc-ssi-unveils-the-aml-an-underground-locating-game-changer.pdf
27 Federal Highway Administration (FHWA). (2024). Underground Utilities – Cable and Pipe Locators – Electrical Line. https://infotechnology.fhwa.dot.gov/cable-and-pipe-locators-electrical-line/
28 Hoffer, T. (2019, April 23). What's the Difference Between Passive vs Active Utility Locating Services? https://www.softdig.com/blog/passive-active-locating-services/
29 Hoffer, T. (2019, April 23). What's the Difference Between Passive vs Active Utility Locating Services? https://www.softdig.com/blog/passive-active-locating-services/

30 GlobalInfoResearch. (2024, April 2). Global Cable and Pipe Locators Supply, Demand and Key Producers, 2024–2030. https://www.globalinforesearch.com/reports/2104393/cable-and-pipe-locators
31 Team EJP. (2024). How to use tracer wire in your water system. https://cdn2.hubspot.net/hubfs/406615/ejp-tracer-wire-offer.pdf
32 Municipal Advisory Board. (2024). MAB Model Specifications for PE4710 Buried Potable Water Service, Distribution, and Transmission Pipes and Fittings. https://plasticpipe.org/common/Uploaded%20files/Technical/MAB-03.pdf
33 Installation of plastic pipe, 49 C.F.R. § 192.321. (2026). Electronic Code of Federal Regulations. https://www.ecfr.gov/current/title-49/subtitle-B/chapter-I/subchapter-D/part-192/subpart-G/section-192.321
34 N.C. Gen. Stat. §§ 87-115 to 87-129 (2025). https://www.ncleg.net/EnactedLegislation/Statutes/HTML/ByArticle/Chapter_87/Article_8A.html
35 Reduct NV. (2022). Gyroscopic pipeline mapping. https://reduct.net/en/gyroscopic-pipeline-mapping
36 GPRS. (2024). How To Protect Your Municipality from the Dangers of Directional Drilling. https://www.gp-radar.com/article/how-to-protect-your-municipality-from-the-dangers-of-directional-drilling
37 National Transportation Safety Board (NTSB). (1976, November 12). Safety Recommendations P-76-83 through P-76-86. https://www.ntsb.gov/safety/safety-recs/recletters/P76_83_86.pdf
38 Call Before You Clear. (2026). Know the Risk of Cross-Bore Encounters. https://callbeforeyouclear.com/
39 Common Ground Alliance (CGA). (2026, February). Best Practices Guide Version 22.0; Chapter 8; 8.11 Cross-Bore Determination and Mitigation. https://bestpractices.commongroundalliance.com/8-Public-Education-and-Awareness/811-Cross-Bore-Determination-and-Mitigation
40 Environmental Protection Agency. (2025, April 18). Ground Penetrating Radar (GPR). https://www.epa.gov/environmental-geophysics/ground-penetrating-radar-gpr
41 USRadar INC. (2022, February 18). What can you detect with ground-penetrating radar? https://usradar.com/blog/what-can-you-detect-with-ground-penetrating-radar
42 The Utility Expo. (2025, May 16). Pros – Cons of a Vacuum Excavator. https://www.theutilityexpo.com/news/pros-cons-of-a-vacuum-excavator
43 Illinois Underground Utility Facilities Damage Prevention Act, 220 Ill. Comp. Stat. 50/2 (2025). https://www.ilga.gov/documents/legislation/ilcs/documents/022000500K2.htm
44 Iadanza, M., & Moorhead, R. (2018, November 28). Understanding The Cost And Value Of Good Public And Private Locates. https://actsnowinc.com/global-811-magazine/understanding-the-cost-and-value-of-good-public-and-private-locates
45 CGA counts unique tickets that reached state 811 centers.
46 Common Ground Alliance. (2024). 2023 DIRT report: Analysis & recommendations. https://dirt.commongroundalliance.com/2023-DIRT-Report

Chapter 3

Contractors Doing the Digging

Arriving at the property, there was not much to go on. A sea of dried out grass seemingly went all the way to the horizon. From his truck, he walked a few paces and pulled out his iPad and a clipboard to double check the property location and description. His eyes scanned the area to look for landmarks like the small shed and an overhead powerline. Walking a bit further, he saw several bright streaks of spray paint.

The closer he got, the more colors revealed themselves. A green line ran under his feet until it reached the street, where he could see that it met a storm drain. A red line ran another path yards away. He could see blue, yellow, and orange taking different routes, with different styles of marking, including what looked to be a logo or initials, a few numbers, and a series of parallel lines with dots in the middle. He had to trust that these markings were correct, accurate, and would keep him safe if he followed the safe digging practices.

This hotbed of markings was fortunately a ways away from where he intended to work. As he walked back to his truck, he took a deep breath, and breathing out into the crisp morning air, a puff of visible steam dissipated as quickly as it had appeared. He walked past the truck and stepped over the trailer hitch, deciding which point to release first on the flatbed before rolling the mini excavator off the platform and onto the dirt.

This was his first job of the year, and he was happy for the work, knowing projects are a bit fewer and farther between in the cold seasons. Taking his seat in the excavator, his hands took over the controls and muscle memory went to work. Before long, he would break ground and clear away mounds of dirt.

As a professional excavator, Randy has seen it all. From being contracted to dig pools at resorts to trenches for farmers, the not-infrequent septic tank installation (and hopefully less frequent future septic tank removals), and plenty of ground breaking for building foundations, he knew his way around power tools and dirt. He even picked up an impressive knowledge of soil types, textures, and ecological insights over the years. What has predominantly changed during his career is technology.

He laughed to himself thinking of his first few jobs in the 1980s as a teenager, when he scrambled to call different utility companies a week before he started work. Then a heartier chuckle – this one even sounding out loud into the cool air – at the thought of pulling out a glass screen on the job site from which he could access the total sum of human knowledge! Much less that such a magic device would be so cheap, accessible, and in virtually everyone's hands across the country.

All the technology changes were for the better as far as he was concerned. Sure there are a few redundancies, but they've made him safer, more efficient, and enabled him to grow his business in incredible ways, collecting professional certifications, attracting new clients, and doing more excavation contracts in a year than he could have scraped together in the 1980s with

DOI: 10.1201/9781003796619-4

a whole neighborhood crew trying to gin up business. The best thing in his mind this morning though, was the white-lining process.

As a responsible excavator, Randy was sure to get the site pre-marked for utility companies and locators well in advance of ever moving earth. The last thing he needed was a locator misunderstanding where he intended to dig or a utility company screening the ticket too quickly.

As Randy would put it, "the old way" to do this was spray painting white lines on site where he planned to dig. He smirked to himself describing things as "old" when they had only just changed, or he was in a leading state where law and practice were innovative. Even his brother, just one state over, was still using the "old way" even though for Randy, this switch was only a year old.

But the *old way* that is still the *current way* in a lot of places is limited to just physical white-lining. That used to mean showing up on site days ahead of the job (when he very often was at another job site or working on administrative tasks to keep the business going). It was a time sink, a physical chore, and had real costs like gas and paint. Randy always did it though.

But not for this job.

He always had consultations with his clients, almost always visited the site multiple times while planning, and made sure he had clear communication with the project manager or landowner through phone calls, emails of maps and plans back and forth, and more. He took the responsibility for every job himself to notify the 811 Center so that nothing fell through the cracks. As technology advanced and this collaboration with all his stakeholders got easier, he switched to white-lining – *electronically*!

Now Randy doesn't visit the site an extra time three or so days before a job to spray paint, he simply heads to the website of the Facility Notification Center and enters a locate request online. In the process, he pulls up an aerial map (like Google Maps) and, zoomed into the jobsite, he draws the excavation plan. That makes his life easier, but also lets the Facility Notification Center with its digital maps and the utilities and locators all see the planned dig area from their own offices. Screening no-conflict tickets is easier, faster, and more accurate, and locators on site can still cross reference the planned dig area as they mark any facilities present.

For Randy, it's a win. He isn't just some oldtimer moving dirt, he's a professional contractor utilizing the latest technology to conduct the safest excavation work possible.

– – – – – – – – – – –

Excavation is done by a wide range of stakeholders including home owners, road builders, fence installation companies, utilities, farmers, maritime operators, and many more. The damage prevention community often refers to excavators, regardless of the specific type, by the shorthand as "contractors." In this chapter, we may use professional excavator and contractor interchangeably, and it highlights an important aspect of these professional heavy-equipment operators and skilled earth movers: their plurality.

There are key differences between *general contractors*, *subcontractors*, *utility-specific crews*, and *independent owner-operators*. Further, as we will emphasize here, is that all contractors may be excavators, but not all excavators are contractors. That is important, because damage prevention focuses on the action – *breaking ground*. In this chapter, we hope that is clear, even while often using the shorthand – *contractors* – to illustrate broader industry dynamics.

A general contractor is the overall project owner. They may be a single person or crew and may complete some or even most of the work themselves, but they coordinate and contract with others to ensure the job is completed within the time, budget, and terms agreed to with a customer or client. In many states, the general contractor cannot contact notification centers on behalf of the entire project, requiring every party breaking ground to individually request a locate, even if they are all working in the same area. Subcontractors are sometimes more specialized or smaller companies helping complete the project one task or discipline at a time. They are hired by a general contractor and may do one type of work or part of the overall scope or segments of a larger project. That may include excavation-specific tasks like breaking up concrete, vacuum excavation, grading, or installing. Each subcontractor must often request their own locate, regardless of whether other subcontractors requested for the same location in the same timeframe.

Utility-specific crews may be contracted directly by a utility owner or even employees working on a contract basis, often armed with detailed facility maps and a plan to build, maintain, or repair a specific line. An independent owner-operator may be just a single person with a mini-excavator who does whole projects by himself or works as a subcontractor. This smaller-scale work fits well within the contractor types and demonstrates how stratified the industry is from larger crews and certified general contractors all the way to individuals and very small businesses. However, anyone who excavates is almost certainly doing so around live utility lines. Whether it be electricity, gas, water, or something else, utility lines are only very rarely shut down during excavation. The crews who excavate to build new structures, replace landscaping, or work on the utilities themselves do not have the luxury of digging somewhere else.

There is a relatively finite number of other stakeholders across the industry. There are utility companies or operators; they could be counted if one had the right phonebook or database. There are fewer large locate companies, with the big third-party locators holding the most market share and then dwindling into far smaller businesses. Each state or region has an 811 Center. State Departments of Transportation match the number of states, insurance companies are a well-defined cohort, and so on with railroads, regulators, and road builders. When it comes to excavators though, it's much harder to count.

The reason "contractor" is a fitting synonym is that individuals, partnerships, and small businesses enter the excavating market every day. A proliferation of equipment rental businesses means just about anyone with a driver's license, truck, and ability to pay can gain access to power augers, mini excavators, and a host of other equipment. Even one of these authors has dabbled in such activity more than once – but in full transparency, in doing so cut his neighbors internet cable, despite notifying the 811 Center and having markings completed. Alleging no fault, many stakeholders were involved, which explains why root causes can be difficult to track down and why a small error or miscommunication anywhere in the process can lead to a near miss or damage.

There are tens of thousands of contract excavators throughout the country, often disconnected from one another and with little organized structure or group power unless they belong to one of the many professional contractor associations. They are small businesses doing their own thing, with a wide range of skill and sophistication. But as a class, the excavation contractor is a profession like any other, with a noble calling, a strong work ethic, and important value to the community.

Who Are They?

Where damage prevention is concerned, contractors are more than just excavators, they are a critical component of ensuring the resilience and longevity of our power, water, gas, communications, and waste infrastructure throughout the country. Their job starts with planning and preparing a work site and only concludes when the dirt is moved without harming themselves or buried facilities, and the property or project owner fulfills their intended purpose or fills the hole back in properly.

In addition to hired contractors, excavators truly do encompass an incredibly vast scope of people. That is actually one of the problems in the damage prevention industry – some people do not think of themselves as excavators or their work as excavation, and do not use the 811 system. Sometimes they simply do not know about the system, like a teenager doing summer odd jobs or an elderly person gardening. Other times, they don't think the 811 process applies to their work, like a rancher driving fence posts into the ground. The specific laws vary from state to state, but it is important to note that virtually every disturbance of the dirt is excavation and that makes everyone disturbing the dirt an excavator, at least as far as the damage prevention process is concerned.

The homeowner or renter planting trees or shrubs, placing a mailbox, building a deck, installing a pool, or more: They are excavators doing excavation.

The landscaping company pulling out flower beds or installing new foliage, hardscapes, churning soil, or laying sod: They are excavators doing excavation.

The farmers and ranchers driving fence posts in, digging for outhouses or septic tanks, burying a horse, or any other imaginable digging at any depth: They are excavators doing excavation.

The dredgers conducting marine restoration, deepening a port facility, sourcing raw materials, or keeping waterways navigable: They are excavators doing excavation.

The corporate-employed professional engineers managing urban bus stop design and building who may lay a new foundation, remove existing structures, or connect a new station to power and other facilities: They are excavators doing excavation.

Even state Departments of Transportation or municipal public works crews building or repairing roadways, rights-of-way, or more: They are excavators doing excavation.

The class of industry professionals who directly undertake or overlap with excavation is vast. Cell tower contractors building the network of towers that dot the countryside and keep us all connected even have an intersection with excavation damage prevention. While these towers are all above ground, and sometimes up to 400 feet tall, every tower has buried cables/conduit to access electric power and to connect to the communications network. Both require new digging and put in place new facilities that become vulnerable to future digs.

This chapter focuses on contractors, but also includes and extends in more peripheral cases to all private or professional earth movers. They come in all shapes and sizes, young to old, rural to urban, and low-skill to high. They are also all intended to follow the same process for digging safely, even though the details of their work will naturally vary, because no matter who is doing the digging, when live utilities are at stake, everyone is at risk.

With seemingly *everyone* being an excavator at some point in their lives, many may wonder why professional contractors keep excavating around live utilities in the first place. Why not simply dig elsewhere? The answer – which is explained in full throughout this book and in other chapters – is primarily that there is so much existing infrastructure already below the surface that every day, more and more excavation takes place near other utility lines. It is impossible to simply stop digging near utilities – thus the necessity of a professional industry dedicated to damage prevention. In fact, many contractors are specifically contracted *to* dig where utilities exist in order to build, maintain, repair, or replace them.

Fred LeSage, a Senior Risk Engineer with AXA XL insurance provider, summarizes well the breadth of stakeholders doing digging during an ESA/ACTS Town Hall:

> We talk about damages that are caused by professional excavators. Well, professional excavators encompass a broad spectrum of folks that are out there. It can be the utilities themselves. It could be a utility contractor who's putting miles of fiber or gas line in the ground

every day. Or it could be a plumber who calls in four tickets a year because once in a while he's got to dig somewhere to connect a line. Or it could be an electrician who the only time he ever has to call in a dig ticket is when he's putting in streetlights and he has to run underground wiring for that. He does two jobs a year where he does that. They fall into that professional excavator category.

We have noted how broad and diverse the class of excavators or contractors truly is and hinted at the dynamic types of work they complete every day across the country. There are also unifying concepts that apply no matter who is digging, no matter where, and no matter the job. These are the requirements, which do vary by state law, but only in the particulars.

Process from Start to Finish

While weeks or months may go into planning behind the scenes, the damage prevention process kicks into high gear about a week before groundbreaking is planned. Depending on state law and what weekends or holidays may be on the horizon, an excavator must generally provide notice to the 811 Center at least three days before the digging starts.

In 18 out of 50 states, the law requires that the excavator pre-mark the site.[1] With the establishment of universal color codes, the spray paint, stakes, and flags in *white* are designated to be used by the excavator.[2] Otherwise known as "white-lining," this pre-mark is to signal to the locators and utility operators exactly where the digging will take place, and if possible, the actual shape and dimensions of the planned excavation. While only around one-third of states require this step, it is recognized as a consensus industry best practice by the Common Ground Alliance (CGA).[3]

Among the many reasons white-lining is viewed as a best practice is that it elevates the clarity in communication and coordination with other stakeholder groups. Any locator or utility personnel arriving on scene can know exactly where the digging will take place, which can help reduce errors like locating the wrong lot or confirming the exact parameters of the size and scope of the request in conjunction with the ticket information. While the locator must identify and mark all known and discernable subsurface utilities in the ticket footprint, white-lining offers more clarity about the location of the work, preventing unnecessary locating where excavation won't take place. Even if it is not required in a state, any locator would be grateful for accurate white markings.

Excavators also benefit from white-lining, especially when multiple parties are involved in the project planning and execution. For hired contractors, it may be a homeowner or project manager who white-lines, helping aid the contractor in knowing exactly where planned digging is requested. For contractors pre-marking the site themselves, this ensures familiarity with the site before the day they break ground. It can also serve as a gut check on the size and scope of the planned dig – if by spray painting, the needed dimensions are clarified, the contractor may realize the digging may be narrower than originally believed or that a wider area is truly in need of excavation, prompting an update to the ticket and a safer process. Indeed, locators likely prefer white-lining so that contractors do have to mark, which would prevent calling in miles of tickets or unreasonably large areas.

There are many positive byproducts of white-lining, but two more worth mentioning briefly, as they impact many different stakeholders and the public.

First is the potential for less "state-sanctioned graffiti" complaints. White-lining reduces the area that will end up with paint and flags being used to identify the facilities. For instance, a company may be planning to install a series of trees as a part of a landscaping project and simply request a locate for the entire lot or the "front yard." This will result in all buried facilities being marked, regardless of where they are in relation to where the trees will be installed. If the project owner simply drew circles with white paint where the trees were being planted, ticket scope could be narrowed and less paint and flags required, reducing what the property owner would see as needless paint all over the area and nowhere near where the trees were being installed.

Second, white-lining can reduce locate times. Locate technicians are frequently pushed to the limit trying to meet state law time requirements. White-lining can save locators a significant amount of time, which helps everyone involved with the project. As technology advances, these benefits are likely to compound with still other advantages.

The advent of electronic white-lining is making the process easier and faster and can now take place as a part of the online locate request. Systems vary from among 811 Centers that offer electronic white-lining, but the end result is that the excavator draws a digital white-line on a map showing their proposed excavation site boundary. Offering this online option can save the excavator a trip to the dig site, and making it part of the online ticket request process increases the likelihood that it will be done. The map offering an aerial view also helps ensure landmarks and total excavation size and shape are accurately designated.

Finally, electronic white-lining can be used in conjunction with physical white-lining, as it has distinct advantages even while carrying similar benefits to the physical marking. Virtual white-lines are already digitally recorded, making it easier for call centers and utilities to match the image to their records. This can improve ticket screening (i.e., the utility owner checking their GIS data and confidently "screening" the ticket out so they do not have to send a locator because their facilities are not in conflict) and ensure all relevant stakeholders have an accessible copy of the planned excavation on their own computer or mobile device. The utility owner can know without having to see the site himself what the plan is, which serves as a multi-step verification of work when the locator then completes his job and sends photographic verification in as well. This layered approach also preserves the dig site in its undisturbed state should any future investigations be needed. In a traditional accident investigation scenario, even if white-lining was completed, all the markings are likely destroyed by excavation. The virtual processes hold them in place and create more clarity, certainty, and improved communication. It also prevents false after-the-fact white-lining by excavators trying to shift liability, which though rare is anecdotally passed around in industry storytelling.

Once the site is pre-marked by the excavator, it is ready for the locators to arrive and use the other universal colors to mark out the site as needed. With each locate request potentially triggering four to six local utilities, the excavator needs clarity about how many companies may need to mark and whether they are all done with their marking by the time the three-day notice matures. In other words, a responsible excavator doesn't put shovel in ground at 7:00 am on dig day just because the clock ran out, they still have to confirm all the utilities that have lines beneath the site have completed their marking. That requires a positive response.

Positive Response Creates a Win-Win Scenario

Positive response is the formal notification from the facility owner or locator to the excavator that the locate request has been received, and that either the lines have been marked, unmarked, or the site has been cleared (no facilities in conflict).[4] This confirmation is normally communicated through the state's 811 Center. In many states, excavators can log in to the 811 Center portal to view responses from all notified utilities – an electronic positive response portal or "ticket check" system. Some states have mandated positive response as a part of their dig laws.[5] Across all states, the definition of positive response varies considerably, with some including minimal language that allows spray paint on the ground to qualify as a positive response and other requiring robust communication to confirm all parties have been fully informed about the dig site.[6]

States often employ positive response codes. These are often two-digit numbers that represent a scenario, such as in Virginia, where 10 means "marked," 30 means "no conflict," and 40 means "agreed to meeting as proposed by excavator." There are codes for virtually every scenario to express more complicated messages and dynamics, including different marking timelines, meeting locations, abandoned lines detected, access to property prevented, private locator required, additional records exchanged or required, extraordinary circumstances existing, and more. Importantly, codes can also change during the life of a project and need to be checked frequently through the website, app, or other communication platform.

While offering a standardized code can simplify, it also adds complexity that contractors have to learn and react to. 811 Centers can also analyze how these codes are being used to conduct data analysis on excavation readiness, stakeholder behavior, and to learn what new messaging and education may be required to ensure safety and efficiency. As they work on electronic platforms to facilitate positive response, these two or three-digit codes (which vary by state in meaning, total number of codes, and how they're used) can be explained or replaced by clear text, accompanied by utility contact information, or even alongside more enhanced information like attachments and notes sections for the clearest positive response possible.

There are many advantages to positive responses. Excavators know definitively if all the utilities have completed their locates before beginning work. This can be especially helpful when a facility owner clears a site (no facilities to mark), because they may not put paint on the ground if there are no facilities there, leaving the excavators to wonder if the missing facility owner is still coming. Locators and facility owners get fewer calls from excavators asking if they have completed the locate. It also creates a comprehensive documentation trail regarding whether all the locates were done on time in case there is damage.

During an ACTS Town Hall titled "*The excavator perspective ... Who's going to get the job done, if we don't do it?*" Roy Sachleben, of Star Construction, explains how positive response can make their job easier and safer:

> If we know a utility is there and we do not have paint on the ground, we do not dig. We tell our crews all the time, if you get out to a job site and you don't see any paint on the ground, it may be a godsend, but it also may be somebody else's mistake, and you're going to have a really bad day if you don't verify that.

Like everything, there are some reasons for not having positive response systems. The positive response does take some staff time and smaller utilities may not have the technology and training to post timely responses. Adding a positive response system does require an upfront investment in software and training, and when the ROI is based on labor savings and/or avoiding damage, it can be hard to track. With that said, some state laws only require a document left at the site or a text, call, email, or other virtual communication, which can be accomplished by the utility staff or locator as a component of completing their marking.

Markings and Locate Tickets Have a Life

Even though a locate request is made and a ticket is issued, the excavator may not be done dealing with the Facility Notification Center, because a locate ticket normally has an expiration date.[7] Tickets often expire after 14 days, but in a few states, there is no limit. The excavator is typically responsible to keep the marks visible for the life of the ticket. If they are still working past the expiration date, they must request another locate, with the same lead time as the original locate request.

Each state or province sets their own ticket life. The length of the ticket life may be impacted by the climate conditions like frequent rain, snow, etc., as well as the level of construction activity. The size of the job matters here. If an excavator calls in a mile of locate requests, but is only going to complete the first quarter mile of work within two weeks, he will have to call in more locates over the course of his project. This strains locators. If the excavator relies on the marks completed originally and does not call in updated tickets as he progresses, he may be in violation of the law, setting up potential fines or liability risks, and increasing the likelihood of a damage due to aging and less up-to-date markings (especially where flags are used, because these can be removed erroneously more readily than paint on the ground).

Above all, the process has multiple steps. It requires engaged stakeholders who know the full process, know the laws they are obligated to abide by, and are willing to collaborate and communicate with the other stakeholder groups throughout the entire project, not just once.

The basic steps for safe digging:

1 *White-line or pre-mark the proposed dig area*

- Excavator also must ensure access to the dig site and that the site is prepared for utility marking (e.g., clearing tall grass and unlocking gates)

2 *Request a locate by calling 811 or using ClickBeforeYouDig.com*
3 *Wait the Required Amount of Time*

- State laws vary, but typically 2–3 business days.

4 *Confirm* – Check the Marks and Positive Response

- Ensure all utilities have responded and marks are clear and accurate.
- Double check for any positive response codes that have changed throughout the project.

5 *Respect the Marks*

- Maintain the marks during your project.

6 *Dig Safely*

- Hand dig or vacuum excavate within the tolerance zone (usually 18–36 inches from each mark or the utility itself once daylighted).
- Be aware of changes in soil color, texture, or density – these may indicate a buried utility.

In the "Defending Utility Damage Claims" ESA Town Hall, Fred LeSage, Senior Risk Engineer, AXA XL, talks about the challenges contractors face when operating in multiple states because the laws vary:

> … it's very important when a contractor moves into a different jurisdiction that you need to understand that there are differences … things like tolerance zones and wait times and whether you get a positive response are all functions of what is required under the 811 law for the jurisdiction where you're working.

Once the site is marked, positive responses are posted and confirmed, and the waiting period is over, contractors can legally begin their work. But breaking ground does not only require those preliminary steps; the next thing the contractor does is observe the markings that may be in or near where they intend to dig and plan for how to complete their work safely.

Respecting Tolerance Zones

Every state has a law designating the tolerance zone – or space around the locators marking representing the buffer zone for the true location of the buried facility. By state laws, these vary between 18 inches and 36 inches. The definition also varies, with some states defining the tolerance zone on either side of the markings (e.g., spray paint or flags) and others defining it on either side of the utility itself, which must be potholed or daylighted to determine precisely. According to the CGA Best Practice guide, the tolerance zone is a distance on *each* side of the outer edge of the pipeline or cable where you must not dig with mechanized equipment. As depicted in the diagram from the previous chapter, if the tolerance zone is 24″ and the pipeline diameter is 2″ that translates into a 50″ area, where no mechanized equipment may be used.

In fact, utilities are usually not tightly clustered or placed in a single conduit. If utilities are near one another but not in the same conduit, they may overlap with one another's tolerance zones, effectively making the tolerance zone the distance to the left of the leftmost and the right of the rightmost facility. This may mean that the no-go zone can turn into the entire sidewalk or property, at which point the excavator proceeds entirely with hand tools or vacuum excavation. While helpful, it can also seemingly defeat part of the purpose of proactive markouts. A similarly counterintuitive concept is that vacuum excavation is allowed inside of a tolerance zone, meaning one could essentially vacuum excavate anywhere because it does not matter if utilities are below, yet vacuum excavation still requires its own locate request – just for someone to mark and the excavator to vacuum excavate right where the marking is anyway!

If the work involves something entirely unrelated to existing or installing of new utility lines, such as planting a tree or digging out a pool or foundation, then respecting the tolerance zone means more than simply not digging where the spray paint or flags are. It means in some cases pulling out a measuring stick and opting for safety over expediency. Many skilled contractors may believe they can navigate the dig site carefully and efficiently with their power tools or heavy equipment. The important thing is that there is natural error and approximation, even with sophisticated technology at play.

Locate technology can, in many cases, pinpoint exact pathways of various facilities, but may also be inaccurate due to tolerances of the tools and methods used, interference, and depth. Then you have the locator applying the spray paint or flags, which may also vary by one or a few inches from the exact path of the buried facility that may be inches or feet below the surface. The tolerance zone is designed for just these small margins as well as the fact that soil composition may lead certain facilities to shift below the surface or several lines may run parallel, widening the risk zone.

When the excavator does begin the work, it may be helpful to have a second observer or spotter on hand to look out for issues. Depending on the depth of the line, and the purpose of the excavation, the person may need to dig directly where the spray paint is or around the tolerance zone (especially if it is work on the facility itself!) so hand digging or potholing, soft or vacuum excavation, or another approved method may be used. The main key is that the excavator in this

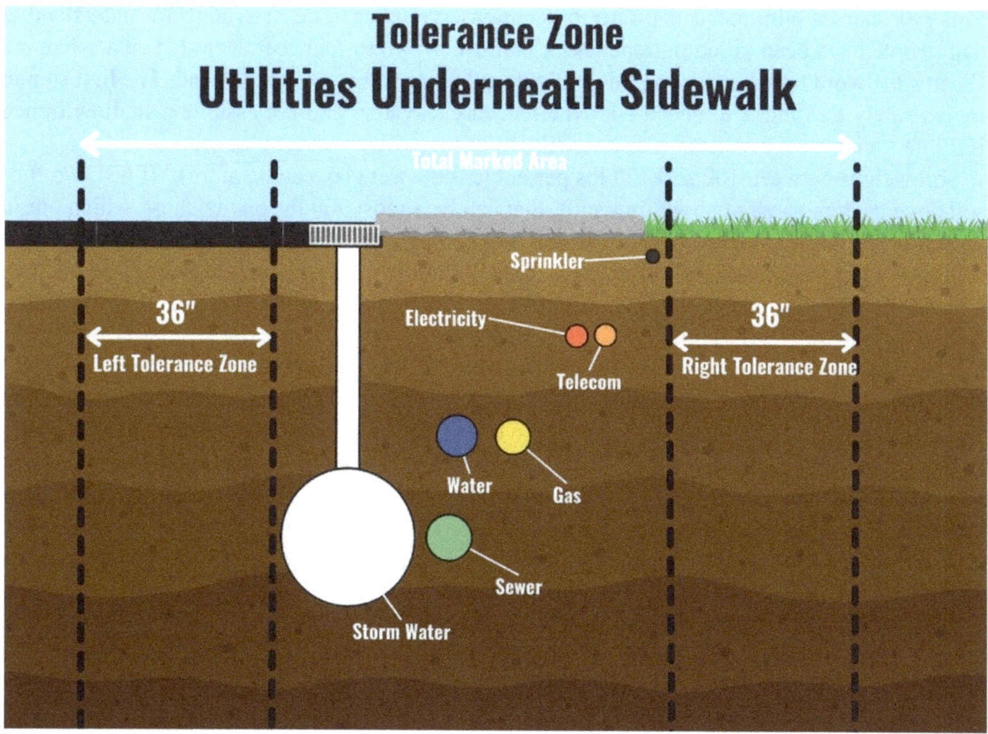

Many utilities in close proximity can create an effective tolerance zone that spans a wider area, necessitating hand digging for an entire project.

zone is not using pointed or mechanized tools like a pickaxe or large machinery like an excavator bucket.

> I know it [potholing] saved us so many times. You just can't trust those marks. They need to be verified.
>
> (Will Waldsmith, Field Supervisor at CST Utilities. *ACTS Town Hall*)

In many cases, contractors are on scene to dig exactly in the tolerance zone! As we have described above, the key distinction is not that digging cannot take place within this zone, but the type of digging cannot involve power tools and large machinery. In reality, contractors may begin a job and switch from mechanical excavation to hand digging and alternative methods in order to maintain existing utility lines or install new ones in an existing corridor. It is often the case, then, that the contractor not only does dig where the spray paint indicates lines exist, but continues until the lines are fully exposed, then protects and maintains them as they complete the job.

Risk and Complexity

The job of an excavator can be complex and risky if safety procedures are not followed. Trench stabilizing is a critical part of excavation safety and is the most likely cause of a fatal accident.[8]

This risk can be eliminated if proper precautions are taken. The first thing to understand is that people have been killed in trenches as shallow as two to four feet deep. If you are like me (Scott), the word trench made me think of a deep hole that was over your head. The first step in trench safety training is to educate inexperienced excavators and not assume a shallow trench is a safe trench.

Soil weighs between 100 and 120 lbs per cubic foot (wet clay can be more).[9] If just two ft of soil covers a torso area of about four sq ft, that can be almost 900 lbs on the chest, which pins a person down, prevents them from breathing and compresses their organs. I have been exposed to trench safety and proper shoring many times including seeing mock resources by the Los Angeles County Fire Department, as well as a rescue demonstration using vacuum excavation in Illinois.

The most powerful lesson for me has been hearing Eric Giguere (Safety Awareness Solutions) speak at several of our conferences over the years. His story is heart wrenching. He married on September 28, 2002 and was scheduled to leave for his honeymoon after his shift when he went to work on Friday, October 4, 2002. Eric had 5 years of experience as a union laborer and was two months into his first trenching project. The crew was installing a 12-inch water main. Eric got down on his knee into the six-and-a-half-foot trench to repair a drainage pipe, and the trench walls collapsed. He was buried under several feet of soil.

Based on the depth and dimensions, Eric experienced, likely over 1,200 pounds (over half a ton) of soil weight would have been holding him down.

As Eric recounts the story, he says he obviously panicked and was unable to breath under the crushing weight. Though, he says that his fear began to subside, and a feeling of warmth and comfort overcame him. Eric was dying.

He regained consciousness in the hospital, where he learned of the quick action of his team to dig him out. The doctors told Eric that he suffered permanent brain damage from the lack of oxygen and that he would likely face a life with permanent disability. Eric fought hard and beat the odds, avoiding permanent disability. Yet, he still shares that it is difficult to be in a dark room, and that his marriage failed because, as he says, he was no longer the person his wife married.

Eric also talks about grappling with the fact that his mistake put all of his co-workers in a horrible position. Knowing that if Eric was not dug out within a very short period of time, he would have had no chance to survive. This forced his co-workers to decide whether to use shovels or an excavator to get to him in time. Both of those options could lead to killing Eric rather than saving him, but there was no chance they could reach him in time by digging by hand.

As Eric says, he is an average guy who had the typical attitude toward safety: "Accidents happen, but not to me," until one nearly took his life and did change it forever. Eric never stopped fighting to beat the odds, and now his mission in life is to make sure the arrogance he had toward safety doesn't become yours. His message is simple and serious, "This can happen to you." Eric has become a powerful speaker with a message that is impossible not to absorb.

OSHA regulations spell out when protective systems (shoring, shielding, or sloping) are required. A link to the OSHA requirements is in the appendix of this chapter.

Another challenge for excavators can be abandoned pipelines and cables. This is covered in more detail in the next chapter, but it is important to note here that if these abandoned lines are not located and marked, they can cause delays and pose safety risks for the contractor. The type of excavation and equipment used may determine the risks of striking other buried lines – including abandoned ones.

Right Tools for the Job

One distinguishing factor between casual diggers and professional contractors is what they're on a site to do. That may also determine the equipment used. Common equipment used to break ground range from simple tools to heavy machinery. Rakes, shovels, augers, and rock bars, are all hand-held but can sever, crack, or weaken pipelines, remove protective corrosion barriers, cables, and wires that may be as shallow as two inches below the surface. When power tools and machinery are employed, the stakes rise in proportion.

Mechanized excavator buckets are far removed from the sensitivity of an excavator's hands. Skilled operators with lots of real world experience know how the dirt feels and how the machinery performs when it contacts dirt, a tree root, or rocks beneath the surface. There are very few government-mandated requirements for training to operate excavation equipment. OSHA does require employers to ensure the people they have operating equipment are competent and trained (per OSHA 1926.21).[10] Most professional contractors have robust training programs to ensure the safety of their team and to avoid costly damages, but there are still times when accidents occur.

Technology will not replace training, but systems have been developed that will alert a backhoe operator if there is something in the dig path or if the bucket is crossing into a hand dig zone.[11] This technology is exciting and could revolutionize the industry. However, even with more advanced technology, well trained, skilled operators are still the best damage prevention tool.[12]

Power digging tools cover a wide range of equipment in addition to the large equipment most people associate with excavation. Mechanized earth disturbing tools include augers, jackhammers, mini-excavators, trenchers, directional drilling, boring, etc. Each of these types of equipment require different training and pose different threats to unmarked buried infrastructure. Most of this equipment is available for rent to anyone with a driver's license, which makes stringent, consistent training a real challenge outside of professional excavators.

Even with hand tools, a pipeline or cable can be damaged. These must be reported to the facility owner. The smallest ding to a pipeline can lead to a leak years down the road and a potential disaster, involving an explosion with injuries or even fatalities.

PHMSA Also Requires Calling 911 if Any Gas Product Is Observed Escaping

Not all excavation involves tearing open a large section of earth with a backhoe, however. Increasingly, there are more minimally invasive options for modern work. Whether for small pipelines or conduit, horizontal directional drilling (HDD) moves laterally beneath the surface rather than making a giant excavation. Installing pipelines and cables under paved surfaces has always been expensive, especially under roads. Think of HDD as tunneling sideways underground without digging a trench. Rather than tearing up a road, sidewalk, or yard to lay a pipe or cable, a special drilling rig bores a small tunnel (called a pilot hole) horizontally from one point to another.[13] Using HDD also keeps roads open with little disruption to above ground activities.

The process starts with the creation of a pilot hole that follows a design bore path. This drill head bores a narrow path to the exit side. A reamer replaces the drill bit, and the drill string is pulled back along the bore path toward the rig creating an enlarged hole to fit the cable or pipe. The pipe or conduit is then pulled back though the drilled path.

There are numerous advantages to HDD, however, when it is performed by a team that is not properly trained, it can cause damage to other buried infrastructure. Getting a locate done is required by law (with exceptions for some utility types) and is especially important with an

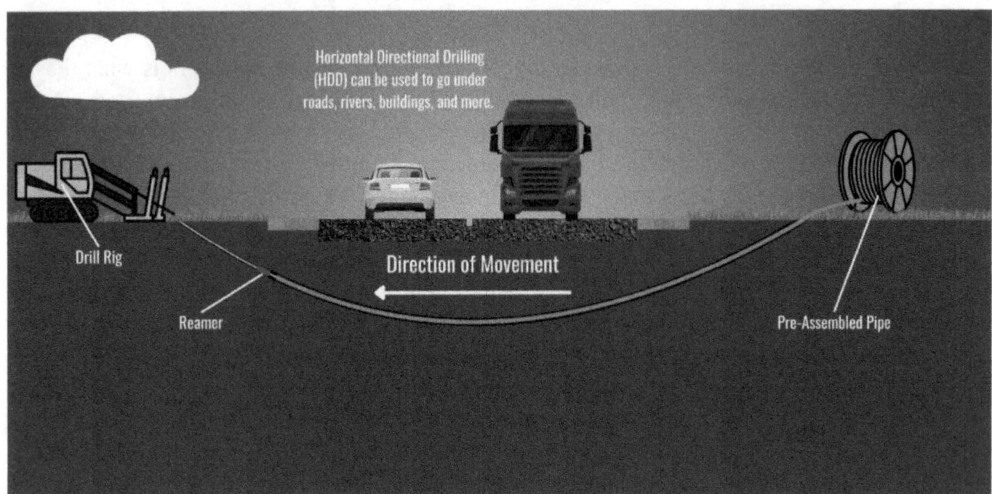

Representation of the horizontal directional drilling process.

HDD project that has other critical infrastructure in the area. In some situations, subsurface utility engineering (SUE) may be needed. Once the drill path is carefully planned and the locates are completed, best practices require that potholes be performed anywhere another pipeline or conduit will be crossed and left open until after installation. A spotter should be able to see the drill head, reamer, and pipe/conduit crossing the other facilities. When done correctly, HDD is a very safe and reliable option, but it requires specialized training and measures.

HDD exploded during the initial fiber installation boom of the 1990s, because massive amounts of fiber could be installed quickly without excavating streets, sidewalks, and yards.[14] Some problems surfaced because the fiber companies were in a race to be the first ones in an area to nail down the customers. This intense demand attracted lots of new people into the HDD business, with no licensing required. Damages became a major issue in many areas and moratoriums were imposed by some cities due to the number of damages to existing underground utilities.[15]

Today, HDD is commonly used for gas, water, sewer and electrical line installation. As another wave of fiber installation is underway, safety practices and training are crucial. Responsible excavators work hand in hand with the utility company to install new infrastructure and protect what is already below. Before we look at why contractors are so invested in avoiding damages, it's worth acknowledging how the industry perceives their role when things go wrong.

Both Perpetrator and Victim

Every excavation damage comes at the tip of an excavator's equipment. This is a truism, as incidents of other types – e.g., corrosion – would not be a type of *excavation* damage (also known as an "outside force" damage). With hundreds of thousands of excavation damage incidents taking place every year in the United States alone, or even over a million as many stakeholders believe, excavators bear the brunt of the blame. But that isn't the whole story.

Sachleben also points out: "Nobody remembers which company came out to do the locating or didn't do the locating a week or 10 days ago or wherever it was. So, it's very easy for us [excavators] to look like the problem."

While they *cause* every damage, not every incident is their *fault*, even if the fingers almost always point at the excavator. Consider industry dynamics as well, that excavators use the system for free. They access the free service of locate requests and utility marks before their work, and may end up receiving only *what they pay for*. This is not to blame other parties, but recognize certain legitimate dynamics.

An excavator could be anyone – someone with a great track record, professional reputation, certifications, and an upstanding citizen, someone suspicious with nefarious intent, someone entirely new to the industry, and everyone in between – and whoever it is, they are not necessarily paying or signing agreements with the utility or other parties. With this broad and ambiguous background at play, many utility companies are not bending over backward to hand out high-quality and accurate digitalized maps of all their assets – many of which may come with legitimate regional or national security concerns if they were used inappropriately. Of course, public markers are placed all across local communities and the locate request process is open to anyone already.

So, the excavator who causes a damage may have been underequipped by another stakeholder group, there could have been honest (read: no malfeasance) miscommunication between parties, mere weather or other third-party could disrupt markings, or far more. Any of these factors could lead an excavator to ultimately strike a utility line and not technically be their fault. With the right framing, the excavator is very often the victim of poor or incomplete information. And even when they are the ones at fault, they are often also the first or only casualties on scene from an incident, making them a victim in a more direct sense.

> If there is something unknown due to poor marks or mismarks, that could be something that falls to the owner of the utility since it's their responsibility to mark it. However, there's an old phrase I used to hear guys say out in the field, 'no paint, no problem.' It's not necessarily the case because even if those marks could be off, if you have prior knowledge, you can still be held liable.
> (Will Waldsmith, Field Supervisor at CST Utilities. *ACTS Town Hall*)

Yet, because the excavator hit the line, they are often blamed as perpetrators. This is a natural response, but once a thorough investigation is completed the facts may redirect the blame, often to more than one party. The damage could be the excavator's fault because they didn't follow safe digging procedures, or it could have been caused by a pipeline or cable being mislocated, or not being located at all. An unidentified abandoned line may have been mistaken for an active line. In other words, just like all news stories, it is best to wait for all the facts before jumping to a conclusion.

During the "What does better communication look like?" ACTS Town Hall, Joel Miller, Spear D Construction says what many contractors feel:

> I think there's a misnomer out there that contractors are just ignoring the law or the regulation because we want to. Well, a lot of that is because we have customers who are without heat, or they're without water, or they're without electricity, and if our locators don't meet those timelines that have been set, and then we have to bear that burden. Well, you got to make a call, and you got to wait until somebody shows up on site.

We will be covering the topic of damage investigation in the next chapter on Utilities and Facilities. The facility owner and the company causing the damage will both normally conduct their own investigation (and you can see how overall direct and indirect costs begin to climb even after a damage has been caused and repaired). If a contract locator was involved, they may also investigate. The focus of all the investigations tends to be on liability, which makes sense, but each investigation into the root cause should also provide information that can be used to reduce the chances of future damages.

Forging Ahead, in spite of the Risks

All stakeholders want to see reduced damages to buried infrastructure, but no stakeholder group has a more vested interest in that than contractors. Whether it is an owner operator or a multi-billion-dollar general contractor, they all want everyone on their team to avoid injuries and go home safe every day. The people actually operating the equipment or doing the hand digging obviously have the most personal interest in digging safely.

Despite being the industry segment with the most skin in the game, contractors have among the least capital. They may be small businesses or individual independent contractors, and while associations and coalitions exist, they are unlikely to invest significant resources into research and development or to design and build new tools. But that has not stopped this stakeholder group from working with other companies and seeking ways to improve safety. Third-party vendors are always innovating – whether it is material sciences to improve pipeline or conduit coating, or technology integration directly into tools for locating or excavating. While adhering to CGA best practices and following the state and local laws is the baseline, still more is often needed due to mistakes, miscommunications, and other variables. That *more* is on the way.

To help incentivize better work on the job site, one company, Crewscope, has developed software to use performance and reward structure to keep teams safe and to work together.[16] The goal is more than just safe and efficient construction work, but more engaged workers and retained talent. An international construction equipment manufacturer has upped technology by incorporating more digital interfaces within the cab of excavators and other machinery.[17] Such work by Magni, and many others, is making construction work not only more intuitive but safer, while streamlining training efforts and improving operational efficiency.

Other vendors have pioneered augmented reality, where a camera pointed at the dig site will provide a virtual representation through the screen of colored lines.[18] This work by vGIS and other companies is not likely to replace spray paint, and relies on accurate records, but is undoubtedly an innovative piece of the evolving puzzle to equip excavators with more resources. Others point to AI as a tool for improving efficiency to modernize excavation and mitigate hazards.[19] Another company has placed locating equipment directly on an excavator bucket, allowing every excavation scoop to alert the equipment operator of the presence and location of any buried lines.[20] This company, RodRadar, believes it will not only make excavators safer, but equip them with more information, which can also be relayed to project owners or utility companies. These and other technologies may serve multiple functions to spin off new innovations and benefits like improved mapping that help the entire ecosystem.

Constant training, use of available technology, and close coordination with all relevant stakeholders is essential, because without contractors out digging every day, the economy would grind to a halt. They understand the risks of the business, yet they do not control all the risks. The risks they face can be both life and death, as well as financial. Even when a contractor follows all the rules and regulations, they cannot eliminate all these risks. Being a contractor requires an enormous amount of skill – and quite frankly bravery – along with knowledge that should never be taken for granted.

Like all professions, there are many entry points. Excavators can get started by renting equipment or going to work for a major corporation. They can get local training or attain multiple certifications. They can use hand tools or high-tech innovation and heavy machinery. And the more they learn about other stakeholder groups, the better they will be at their job.

From a business perspective the contractor, big or small, wants to avoid damages. The spotlight always shines on them when there is damage, because they were the ones doing the digging. Assuming nobody is hurt, the focus quickly turns to the liability for all the repair costs and related costs because most of those are hard costs that someone writes a check to cover. Unfortunately, when the system is not working and locates are late, preventing a contractor from working, the contractor's lost revenue and costs that cannot be recouped, do not get the same attention. This trend seems to be changing, but all stakeholders' needs must be met as a part of the process, rather than only reacting when a damage occurs.

Associations and Their Impact

The last line of defense are well educated contractors. These professional contractor organizations vary, but in most cases, they provide training or training materials. In my experience (Scott), contractors who are active in any of these associations tend to be more professional and take all forms of safety even more seriously. There are no doubt extremely professional safety conscious contractors that do not belong to any association, but I believe they are the exception, not the norm. A focus on damage prevention, excavation safety, and participation with all stakeholder groups is the trend. A great example is the fact that the Minnesota Utility Contractors Association (MUCA) recently rebranded to DIGIN Midwest with a tag line: "The voice of unity in the underground utility industry."[21] They are definitely very forward looking and thinking about the big picture.

As we discussed in the beginning of the chapter, there are many different types of excavators as well as a wide range of specialized contractors. The "contractor" industry has developed a wide range of professional trade associations to share knowledge and promote their members. These are some of the major national organizations in the United States that are involved in damage prevention. Most of these organizations have state or regional chapters which is where most of the day-to-day involvement takes place. While we mention several here, including respective local organization chapters, there are hundreds more.

- *American Fence Association (AFA)*: There are an estimated 70–80 million holes dug for fence posts every year, and fences are often very close to the utility ROW. This combination of volume, hole depth, and location means that the AFA is very interested in damage prevention. www.americanfenceassociation.com
- *American Pipeline Contractors Association (APCA)*: With over 150 member companies, this group represents merit shop pipeline and station contractors and has since 1971. https://www.americanpipeline.org
- *American Road & Transportation Builders Association (ARTBA)*: Their members are primarily involved in large projects and every one involved working around buried infrastructure. artba.org
- *Associated General Contractors of America (AGC)*: This is the largest contractor association with over 27,000 member firms. Virtually all their members are involved in excavation so they have extensive training on safety and are very concerned about damage prevention. www.AGC.org
- *Distribution Contractors Association (DCA)*: Their members install and maintain natural gas distribution pipelines and related underground utility systems, so they have a very specialized interest in damage prevention. www.dcaweb.org
- *National Association of Landscape Professionals (NALP)*: Almost everything their members do involves digging, even though they may not be huge projects. Education is one of their main objectives, especially at the grass roots technician level, so damage prevention is important to them. www.landscapeprofessionals.org

- *National Utility Contractors Association (NUCA)*: Since their members focus on utility construction, including water, sewer, gas, electric, telecom, they also have a big interest in damage prevention and they provide a significant amount of training on the subject. www.nuca.com
- *Pipe Line Contractors Association (PLCA)*: Since their member contractors build and maintain natural gas, crude oil, petroleum products, and other hazardous liquids pipelines, damage prevention is in their DNA. www.plca.org
- *Power and Communication Contractors Association (PCCA)*: These 240 member companies include contractors building broadband systems and electric power facilities, including transmission and distribution lines and substations. Many of these systems are below the ground, while overhead damage prevention also cannot be forgotten. https://www.pccaweb.org/

In addition to these, more specialized associations also exist including, but not limited to: American Subcontractors Association (ASA), Associated Builders and Contractors (ABC), and more niche even including Dredging Contractors of America (DCA), The Pile Driving Contractors Association (PDCA), and so on. We highlighted several, but please forgive any notable omission – we encourage readers to look for even more.

The existence of these organizations is vital for information sharing, workshops and trainings for fellow contractors, pooling resources, educating lawmakers, and even basic solidarity. Not every contractor is a member of an association. The plurality of excavators detailed in this chapter helps explain why that wouldn't be practical, but also explains the value in such organizations being an option.

The damage prevention starts and ends with the contractor. They make the first marks in white and initiate the 811 process by requesting a free locate. That kicks the entire ecosystem into gear, with digital platforms, emails, and other communications moving at the speed of light between stakeholders across the industry. Once the site is excavation-ready, the contractor begins and uses their skill and experience, along with best practices and required procedures, to break ground. Only when they complete their work is *damage prevented,* and the hole or trench can be backfilled or the next phase of work can begin. All of this makes contractors a unique stakeholder group – and a lightning rod for anything that may go wrong.

Bearing all of this on their shoulders, it is vital that responsible contractors be praised for their hard work and those with more to learn do so with the spirit of humility and eagerness to improve. Contractors may not pay when they use the system to request a locate, but they have skin in the game and they hold back disaster more often than they have a hand in creating it.

Chapter 3 Resources

The CGA's Tips for Excavators

The Communications Infrastructure Contractors Association (NATE) 2025 safe digging video designed specifically for their members building cell towers.

5 Steps to Safer Digging video created by the CGA in 2011 – A 10-minute video overview of the key steps, with a U.S. focus.

Excavation Safety Guide – This reference guide is packed with articles and reference materials to help you dig safely.

CGA Training includes modules on many key excavation safety topics.

ESA/ACTS Now Town Halls

The Digging Dangers videos section of this Planet Underground TV – YouTube channel is especially relevant for contractors.

OSHA Regulations and Guidelines

Safety Awareness Solutions

The Dig Daily Dose Newsletter

Notes

1 Dierker, B., & Rogers, O. (2024, September). 2024 Damage Prevention Report Card. https://www.aii.org/wp-content/uploads/2024/10/2024-Damage-Prevention-Report-Card.pdf
2 American Public Works Association (APWA). (2025). Uniform-color-code-1.pdf. https://www.apwa.org/wp-content/uploads/Uniform-Color-Code-1.pdf

3 Common Ground Alliance (CGA). (2026, February). Best Practices Guide Version 22.0; Chapter 5; 5.2 Delineate Area of Proposed Excavation. https://bestpractices.commongroundalliance.com/5-Excavation/502-Delineate-Area-of-Proposed-Excavation
4 Lewis, M. (2023). Positive response. https://digsafelyoregon.com/resources/positive-response/
5 Fla. Stat. § 556.105 (2025). https://www.leg.state.fl.us/statutes/index.cfm?App_mode=Display_Statute&URL=0500-0599/0556/Sections/0556.105.html
6 Dierker, B., & Rogers, O. (2024, September). 2024 Damage Prevention Report Card. https://www.aii.org/wp-content/uploads/2024/10/2024-Damage-Prevention-Report-Card.pdf
7 Alabama 811. (2022, September 20). Life of a locate ticket. https://al811.com/life-of-ticket/
8 Occupational Safety and Health Administration (OSHA). (2015). Trenching and Excavation Safety. https://www.osha.gov/sites/default/files/publications/osha2226.pdf
9 Missouri Department of Labor. (2019). Toolbox talk: Excavation safety. https://labor.mo.gov/media/28016/download
10 Safety training and education, 29 C.F.R. § 1926.21 (2026). https://www.ecfr.gov/current/title-29/subtitle-B/chapter-XVII/part-1926/subpart-C/section-1926.21
11 Leica Geosystems. (2023). Safety Awareness Solutions. https://leica-geosystems.com/products/machine-control-systems/awareness-solutions
12 Occupational Safety and Health Administration (OSHA). (2025). ETool : Construction – trenching and excavations – competent person. https://www.osha.gov/etools/construction/trenching/competent-person
13 Enbridge. (2021). Horizontal Directional Drilling. https://www.enbridge.com/~/media/Enb/Documents/Factsheets/FS_Horizontal_Directional_Drilling.pdf
14 Carpenter, R. (2015, June). HDD in Modern Times. https://undergroundinfrastructure.com/magazine/2015/june-2015-vol-70-no6/features/hdd-in-modern-times
15 Town of Addison, Texas, Ordinance No. 097-039 (1997). https://gis.addisontx.gov/GIS_Documents/Ordinances/1997/097-039%20-%201997.pdf
16 CrewScope. (2026). Real World Applications. https://crewscope.com/use-cases
17 Magni Telescopic Handlers. (2025, October 31). Azienda Magni: The Future at Every Height. https://www.magnith.com/azienda/
18 vSite. (2026, January 15). Augmented reality for utilities and Civil Construction: vSite. https://www.vgis.io/augmented-reality-construction-subsurface-utilities-bim-gis-sue/
19 Bansal, A. (2024, September 26). How AI-Driven Excavation Leads to Faster, Safer Infrastructure Projects. https://www.constructionbusinessowner.com/technology/how-ai-driven-excavation-leads-faster-safer-infrastructure-projects
20 RodRadar. (2025). LDR Excavate. https://rodradar.com/product-ldr-excavate/
21 DIGIN Midwest. (2025). About us – DIGIN Midwest. https://www.diginmidwest.org/page/AboutUs

Chapter 4

Utilities and Facilities

Sarah Lynn woke up early as though it was Christmas morning. Her parents had been preparing for the neighborhood barbecue and cookout for a whole week. Even though it wouldn't start for another six hours, her little mind couldn't bear to stay in bed or even in her room – she had to get downstairs and outside and start preparing for all the games she would play with her friends in the cul-de-sac.

Opening the door at 7:30 am, she could hardly contain her excitement. Maybe it was Christmas after all! Or at least, *Christmas in July!*

To her, Santa must have come this week to prepare her house for the Fourth of July! While his elves bring garnish and streamers for Christmas, this year, they painted the fireworks right on her front lawn! What a beautiful sight! She squealed so loudly, her mom ran out of the kitchen to the front door.

"Sarah Lynn, what is it! Are you alright?" her mom said with breathless urgency.

"Momma, look at all the colors!" Sarah Lynn shot back loudly without regard for waking up any neighbors. "The fireworks came early!"

Laid out in front of her were lines of spray paint in blues, greens, oranges, reds, yellows, and more colors. Just like the diamonds of exploding light for Independence Day, her very own lawn was a dazzling array of color. And it didn't stop in her lawn either.

The sidewalk, her driveway, and the street all had colorful lines and patterns.

Seeing that Sarah Lynn was safe, that another neighborhood friend was emerging down the street, and knowing the meals would not prepare themselves, her mother returned inside with an understated reminder *not to get up to anything.*

Her friend Kayla ran to meet her in the street. Together, Sarah Lynn and Kayla traced the colorful lines all throughout the yard. They sang songs about the fireworks, shared hopes for how fun the day would be, and wondered what all the spray paint really meant. As they traced the lines, one ended up leading to a storm drain, where they also saw a picture of a frog, another dead-ended at the house. One ran straight into a green box near the sidewalk with a giant lightning bolt on it. The girls were thrilled to find another color, this time not spray paint at all, near the side yard.

Just as their neighborhood was lined with flags in patriotic anticipation for July 4th, the house had its own miniature flag parade. These were small orange flags that ran along the yard between Sarah Lynn's house and a neighbor's.

Kayla wanted to pick up a flag and run through the street waving it with pride, but Sarah Lynn said they better not. Someone must have put it there on purpose, and they shouldn't move them.

DOI: 10.1201/9781003796619-5

Kayla was disappointed, but Sarah Lynn proposed they make it a game to find out what they all mean. She started by reading the words on the orange flag:

Underground Telephone Line
Contact 811 Before Digging

"I thought phones were wireless," says Kayla curiously. "What does it mean there are phone lines underground?"

Before Sarah Lynn could even guess at an answer, Kayla ran across the yard, spotting a series of yellow flags! These said:

EVERTON energy
WARNING BURIED GAS LINE
HAND DIG TO LOCATE BEFORE MECHANICAL EQUIPMENT IS USED

The girls pictured gas station gas and were even more confused than ever! Just then, Sarah Lynn's dad called them in for breakfast. Already it was getting hot, as it was July … and in Texas …

Once inside, the girls asked her dad about the spray paint. He told them, "well, you feel that cool AC and see that nice glass of water? It's all thanks to those spray paint lines." He enjoyed the moment of confusion on their faces before explaining that all the things they use and depend on – and take for granted – come to their homes and neighborhood through buried utility lines. Those flags and spray paint markings are to keep the services they depend on going and prevent damage to the gas, cables, and wires when he and Sarah Lynn's uncle redo their lawn later in the week.

– – – – – – – – – – –

Modern Roots

We may take for granted how interconnected modern life is sometimes. While it is difficult to conceptualize, a residential home is physically connected to a power plant, water treatment plant, natural gas storage tanks, and more.

The light switch on the bedroom wall is powered by wires running through the walls that connect to an overhead service drop or buried service lateral. These in turn connect directly to electrical distribution lines, which further upstream connect to high-voltage transmission lines. These originate further back where they are wired into power generation plants. Along the way are substations, voltage increasing and decreasing equipment, and transformers, but at every stage, the two ends remain physically connected.

This same connection is found between a home faucet and water treatment facility, between a stove top and a gas processing facility, a home internet modem and a data center. These physical links are mostly unknown to us, because they are out of sight and out of mind. And, because of the large distances and complex connections, many may fail to think about the infrastructure even when it is common sense that water, power, and gas must be moved through a physical medium and thus be connected with pipes, cables, and wires.

A modern home is like kids playing by talking to each other through tin cans and string. The game only works because the cans are connected to one another. Instead of voice being carried through vibrations, it is all the services we rely on. And if that string is cut, the home is cut off from services.

All of these connections are commonly referred to as utilities. Other names include service lines, facilities, buried lines, laterals, mains, and more. Utilities can also reference the company owning or operating these assets. These are sometimes public companies, municipally owned, or private entities.

Any given home or business is likely to be connected both underground and aerially to multiple utilities. The overhead lines connect to the building with a service drop (a small cable serving just one home) from a distribution network of electrical power. Telephone and datacom may also connect this way. Below ground, laterals (small diameter pipeline connecting one house to a mainline) are becoming increasingly popular, but virtually every building has water and sewage connections, with natural gas being very common.

In effect, an average home likely has four to six utilities connected to it: electric power, natural gas, water, sewage, phone, and internet. Some of these may be provided by the same company, such as a power company providing both electric and gas, or a municipal water service handling both the clean water in and the waste water out of a home.

While not fully comprehensive, a list of common buried pipes, cables, and wires includes:

- Upstream gathering lines – crude oil, natural gas
- Midstream transmission pipelines – crude oil, refined products, natural gas, carbon dioxide, hydrogen
- Downstream distribution pipelines – natural gas, propane, synthetic gas
- Liquefied petroleum gas (LPG) lines
- Natural gas liquids (NGL) lines – butane, ethane, pentane
- CO_2 pipelines – for enhanced oil recovery or carbon capture and storage
- Hydrogen pipelines
- Potable water mains – pressurized clean water for residential/commercial use
- Service lines – smaller offshoots from mains to individual buildings
- Irrigation lines – agriculture, landscaping
- Fire suppression lines – hydrants, sprinklers
- Sanitary sewer lines – gravity-fed or pressurized
- Storm drains/storm sewer lines
- Greywater lines – reused non-potable water systems
- Industrial process water pipelines
- Slurry pipelines – mining, construction material transport
- Brine lines – used in oil/gas and desalination
- Chemical pipelines – ammonia, acids, solvents
- High voltage underground transmission lines
- Medium voltage subtransmission lines
- Primary electrical distribution lines (medium voltage, e.g., 4 kV to 35 kV)
- Secondary electrical distribution lines (low voltage, <1 kV)
- Electrical service drops/laterals
- Private electrical lines – landscape lighting, gate/motor power feeds, outbuilding feeds
- Fiber optic cables – long-haul, middle mile, last mile
- Coaxial and hybrid fiber-coaxial cables – cable TV, older internet infrastructure
- Twisted pair copper wires – telephone, DSL lines
- Cellular network fiber backhaul
- Campus data networks – universities, industrial sites

- Security system cables – CCTV, alarm wiring
- Building automation system cables
- SCADA and control signal lines – pipelines, utilities, substations
- Railroad signal and communication cables
- Traffic signal and streetlight control wiring
- Pipeline cathodic protection wires
- Military communication lines
- Private water lines
- Private sewer or septic system lines
- Private electrical circuits
- Geothermal ground loops
- Low-voltage lighting systems
- Gas grill/firepit lines (natural gas or propane)
- Private fire mains
- Private data networks
- Grounding/earthing electrodes
- Solar panel conduits and cables
- EV charging station feeds
- District heating and cooling lines – steam, chilled water
- Subsurface vacuum systems – waste removal
- Underground pneumatic tubes – bank drive-thrus, hospitals
- Cryogenic pipelines – LNG facilities
- Compressed air lines – industrial campuses
- Deicing fluid or glycol lines – airports

Exactly how many of these exist, their linear mileage, and their proportion are virtually impossible to know. For years, it has been common in the industry and in educational and awareness campaigns to state that there are 20 million miles of underground utilities.[1] Many readers may be surprised to learn this figure originated in another century.

The origin of the 20 million mile statistic can be found in a 1994 workshop hosted by the National Transportation Safety Board (NTSB).[2] Compiled by Bell Communication Research, this figure was an estimate of all buried pipes, cables, and wires in the United States at that time. Just five years later in the 1999 Common Ground Study, this figure was put forward once more with this important caveat: "This figure has likely dramatically increased in the years since."[3]

Yet, over 30 years later, we still hear 20 million being used.[4] And there are good reasons for it – it's a hard number to estimate!

Who is best positioned to have this data? Government agencies may have some records, but are more likely to track incidents, total service capacity, or other metrics, not linear mileage. Moreover, a majority of critical infrastructure in the United States is owned and operated by private companies. Trade associations may have some data but are likewise not focused on the distance each member company spans with their infrastructure.

The 811 Centers seem like the best aggregating source for this data, but consider that their mapping platforms have three key limitations: first is the use of grids and polygons. In many systems, a grid is used, where a utility operator can effectively check a box to indicate they have some infrastructure in that geographic square – and to notify them if excavation is happening there. It doesn't say much about the linear mileage of the facility. Likewise, polygons may simply capture bulk geometric shapes or include buffers indicating it is not a precise representation of the location. What is registered with the 811 Center is known as a notification area, making it intentionally broad and its purpose limited to notifying utility operators of activity in the area near their underground facility.

The second limitation is the shape or route of the buried pipe, cable, or wire. A mile of wire shaped like an "S" could fit all within one grid or polygon. The call center may not even know the shape, and if they did, could not add the linear length of the grid or polygon to get an accurate measure. They also wouldn't spend their time capturing this kind of metric, because for all intents and purposes, it has no value to their operation and mission.

Third, call centers have information on public utilities and those required by state law to be members of the call center, but may not have information on private lines (including those behind the meter) or abandoned lines.

Another option is to contact every utility company in the entire country, which is prohibitive and even then, they may not have this kind of data. How might people consider conduit, through which many individual utility lines run, but are overlapping and concurrent? Who has the time, energy, know-how, interest, and reason to conduct this research? Very few. And that all helps explain why we still say 20 million miles today – it is a huge number, and it gets the job done. People hear it and think, *that's a lot!*

But it is out of date. It was out of date in 1999. And with the advent of the internet and its accompanying infrastructure largely occurring in this century, it is out of date by a wide margin. The population has grown by 80 million people.[5] Around 36 million additional housing units have been built since then.[6,7] Electric power demand has grown by over a trillion kWh in the meantime.[8] According to PHMSA there have been about 730,000 new miles of regulated pipelines installed between 2000 and 2024.[9] Millions of miles of new fiber optic lines have been installed during this time, and more is planned to continue its rapid expansion.[10] New power, water, sewer, and more are added for every new construction, and with over 30 years elapsing, we know these have added up, even if no one was exactly keeping count.

So, the time has come to *bury* 20 million miles.

In its place, the new national benchmark estimate is that there are around 50 million miles of buried utilities in the United States by 2025.[11]

Recalibrating the Over-Under

There is a name for placing new lines below the surface or relocating overhead lines to the subterranean zone. It is called "undergrounding." Some industry voices have espoused a goal to see fully 50 percent of certain utility lines underground by 2040. If that takes place, the multimillion mile figure estimated today is certain to increase in the years ahead. Likewise, the pace of fiber optic deployment alone will materially change the mileage in only a few short years extending to 2030 and beyond, as programs like the Broadband Equity Access and Deployment (BEAD) program are fully funded and dispersed.

Some may ask "why is undergrounding preferred." A visual response may answer that "why" with a "Y."

Overhead lines face their own risks and have unique and recurring maintenance requirements. It can be costly to constantly monitor and trim trees, while it can leave a blight on the area – not to mention weather-related impacts like ice and wind damage compounded by vegetation. In fact, recent data indicate that utilities spend $6 billion to $8 billion annually clearing vegetation

It is common to cut trees into a "Y" shape to give wide clearance for overhead lines. This aesthetically unappealing solution also requires continuous maintenance.

from around overhead lines.[12] In North America, this is estimated to consume the largest single line item in operation and maintenance (O&M) expenses for many utilities. Many newer neighborhoods are built with underground service laterals as a modern default, while older neighborhoods, like the one pictured, often still have overhead lines or are only moved underground after long-term planning and in coordination with other major projects – like other utility installations, new sidewalk building, or road work.

All of this leads to the key point: there is a world beneath our feet wherever we go in the United States. In virtually every corner of the country – and around the globe – something is pulsing with information, water, power, gas, or waste underneath the surface. That is why damage prevention is so important, and that is why the men and women in this profession are constantly holding back disaster.

Utilities – A Who and a What

Similar to our use of "contractors" in the previous chapter as a shorthand for excavators, there is another definition to cover here: utilities.

Very often "utilities" refer to the physical lines in the ground, but may also refer to a class of companies, or a specific company. This can be very confusing. Ultimately, utilities are a who and a what. A public or private utility company (the who) owns or operates the service lines (the what). We hope context will bear out our meaning, but readers should stay on their toes to spot how the same word may be used for one or the other meaning in this book.

Public utilities often include water, sewer, gas, electricity, and now even fiber systems. Tribal Utility Authorities ("Tribes" hereafter) can be similar to municipalities because they often own and operate many utilities. Other facility types are not considered utilities, like telecom lines or gas and oil pipelines. Yet, for our purposes, they are encompassed in the general discussion of "utilities." Broadly defined, we mean the owners and operators of our infrastructure that's buried and the infrastructure itself which the process is designed to protect from damage. In the same way "contractor" was employed as a short hand for "anyone digging" previously, please read "utilities" as any and all lines under the surface when discussing damage prevention, unless specified.

With the "who" depending in part on the "what" – which will also help inform the "where" – the next question may be: what exactly is it that is below our feet. The list above explained many of the public utilities, private lines, and other facilities. Below, we discuss just how some of those work and where you'll find them.

Common Depth Information

In theory, the typical depth of most buried cables and pipelines should range from 12 inches to 48 inches. However, many water, sewer, and storm drain pipelines work based on gravity and they are more commonly three to 13-feet deep, but can be over 100-feet deep in rare circumstances. Then there are critical but less considered facilities that are both submerged and buried – that is, below the surface of the water and the lake or ocean floor. Many tens of thousands of miles of these exist, and often require special 811 locate requests. Their depth also varies as water level and cover can change due to natural events like tides or marine commerce like anchor dragging or dredging.

While some facilities are found deep beneath the surface (including the double surface of marine environments), not all are intended to stay far below. The flip side of that is the fact that anyone who has been in the industry for a few years has seen plenty of service drops in yards that are partially exposed, even without any digging. The depth of pipelines and cables varies dramatically based on the type of pipeline or cable, how deep the frost gets in the area, the soil conditions, whether it is an urban or rural setting, what is on the ground above the facility (a lawn, road, railroad, etc.), and how long ago it was installed.

It is also possible for depth to change over time due to soil conditions, additional excavation, geologic events, weather conditions, and even changes in pressure or effects of the contents of the lines. This inconsistency means that it is not smart to assume anything regarding facilities depth below the paint. Collecting dust and dirt may mean a facility is several feet deeper than it should be, while high erosion can mean a pipe that was originally installed four feet deep is nearly exposed.

Scott cites one example of a pipeline operator that requested marker posts that changed color every six inches. The bottom portion was red to signify danger. Their pipeline was in a right-of-way in a rural setting, where the road was graded regularly, which reduced the cover over their pipeline. They routinely monitored the cover depth, with the different colors on the marker posts providing a very quick visual way to know if the cover was sufficient. If they saw red on the marker post, they knew they needed to take action. Of course, they also had warning decals on the posts as well.

Along with these come different risks and also different types of protection. For instance, some lines are found in conduit, others are by themselves, others have added concrete encasement, and more. In some cases, fiber optic cable was installed within abandoned pipelines.

In the 1990s and 2000s, communications cables were even installed in active water pipelines to avoid the costs associated with digging new trenches or boring new conduits through congested urban areas.[13] This also provided protection for the cables from third-party damage. The cables had to be NSF-61 certified (safe for contact with potable water) and use non-toxic lubricants. However, even with all these safeguards, this practice is not allowed in potable water pipelines in most jurisdictions.

This wide range of burial depths also makes it harder to use horizontal directional drilling (HDD) in some extremely crowded right of ways. In the early days of HDD, some inexperienced operators thought all they had to do was go five feet or deeper and they would go under everything. Some learned the hard way about the diverse depth of some utilities. That resulted in cross-bores, where two pipelines, conduits, or other lines – you might say *share* – the same location. In other words, drilling straight through a live pipeline. Technology and training have dramatically reduced these occurrences with experienced HDD operators.[14]

A reason cross boring is so dangerous is the contents and pressure of many lines. It is also important to note that it can happen because of incomplete or inaccurate mapping, along with some laterals not being locatable without special equipment. In many cases, facility depths are legally required and factor in the public safety risks, pipe or cable material, the contents of the lines, or permanence of the infrastructure. Hazardous liquid and gas pipelines will be far further underground than electric and telecommunication lines.

The depth of the pipelines makes them more costly to install and repair, but additional depth and dirt also serve as one of the protective elements. Electric and telecommunications are not cheap and can be dangerous if hit, but are simpler to repair or replace. From shallowest to deepest, here are some additional things to know about the who, the what, the where, the why, and even the how.

Telecommunications lines, including phone and internet or broadband lines, are made primarily of copper or fiber optics cables, surrounded by protective layers of cladding. While copper is used to transmit electrical signals, fiber optic lines are ultrathin and pure strands of glass that send data as light pulses. Fiber has become the favored option for data and communications

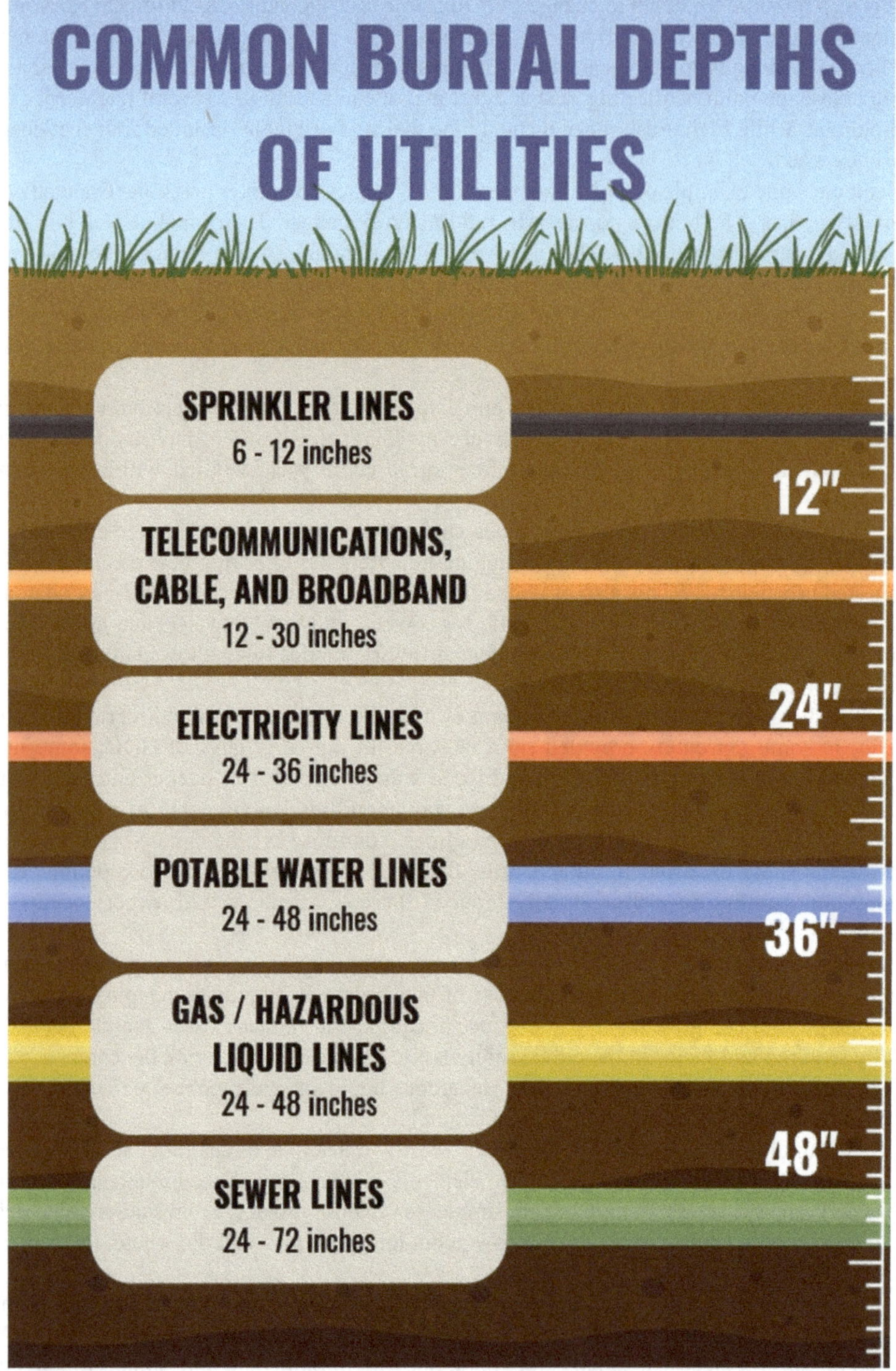

All depths represented are general approximations, which should not be relied upon when digging. These are presented here to provide broad context to readers, but there will be significant variation for any individual line in the real world.

applications because of its far higher bandwidth and speed of transmission.[15] For both copper and fiber, many dozens or hundreds of individual wires or strands are often bundled together in a single "cable" at the long or middle mile distance, where the depth may be around 24 to 48 inches. While these go by many names, including long-haul, backbone, and intercity, in other countries, they bear different names. However, there are commonalities with regard to the importance and sensitivity of these lines.

> In Australia these used to be referred to as trunk or 'inter exchange' and now more commonly inter city or interstate routes. These are the most expensive to repair as there may be a hesitancy to introduce new joints in the fibre cable, creating additional transmission losses. Sometimes the terms main and distribution can apply to telco networks also.
>
> (Chris Ross)

A cable connecting directly to a home, or a service line (lateral), likely includes far fewer strands and may be found closer to 12 inches, or less, below the surface. This is partly why telecommunications are the most struck facility, because they are shallow buried and in places like homes and businesses, where people may dig or do yard maintenance and landscaping without thinking about what is just under the lawn. Still, the consequences of these strikes can be among the lowest stakes.

By and large, high-voltage transmission lines remain overhead, with only a small fraction buried in the United States.[16] The majority of buried electrical infrastructure are distribution and service lines, at lower voltage providing alternative current (AC) power straight into homes and businesses. These may be public, private, or municipal utility companies and are often different from the company owning or operating the long distance transmission lines and towers. The electric lines are most often around two feet below the surface, but depth will in part be determined by whether it is buried directly (e.g., an underground feeder or UF) or in a metal, PVC, or other conduit.

Like with electric lines, natural gas and hazardous liquid pipelines come in at least two major types: transmission and distribution. Of course further upstream, there are also gathering lines, which are often further from population centers. Transmission lines carry large volumes at high pressure for long distances. Distribution likes are far smaller-diameter, low pressure and bring gas or liquids into homes, businesses, manufacturing operations, or other industrial facilities. Gas and petroleum transmission pipelines are usually made from steel, around 24 to 36 inches (but up to larger 42 inches) in diameter, and are pressurized at 200 to 1,500 pounds per square inch.[17] Striking these with heavy machinery can release not only volatile material but highly pressurized gas.

These traverse long distances, often in more rural settings, and commonly pass under farms and ranches and along roads, railways, and electric corridors. They are typically three to six feet below ground.

Distribution lines contain similarly dangerous payloads but in smaller scales, like corrosion resistant plastic and polymer pipelines. While mains may range from two to 24 inches at up to 10 psi, the service lines connected to home meters are often smaller, ranging from a half inch to an inch and a half, at around 0.25 psi.[18] These are more common in urban and suburban areas. As with fiber and electric, gas distribution lines may be nearer to the surface when they get closer to a home or business. The main network is usually three to four feet below the surface, while service lines may be around 12 inches to two feet underground.

Water pipelines are similarly ubiquitous. Everyone everywhere needs potable (or safe) water. From household drinking and bathroom water supply to fire hydrants and sprinkler systems, water connections can be found in every urban, suburban, rural, commercial, and industrial setting. Water isn't itself volatile, but it is pressurized and lines often contain incredible volumes.

Headlines nationwide in recent years have fixated on a national effort to replace older lead pipelines.[19] Newer water pipelines are plastic or polymer-based. The size varies as with other utilities based on the specific type and function, with larger city mains often around six inches to 18 inches, while service lines may be as small as a half inch and up to three or four inches.[20] When it comes to burial depth, water may follow suit with other utilities having shallower depths for service lines and deeper for the larger municipal main system. Unlike other utilities, however, water usually has a minimum allowable depth, so rather than measuring from the surface it is measured from the frost line.[21] This typically places it two to three feet below ground level, and may be deeper at higher latitudes and shallower in warmer parts of the country.

Some view the risk of hitting a water line to be less significant than striking a gas pipeline or a high-voltage cable. In some respects this is true, but water damage can cause significant impacts, including loss of life, and be extremely costly to repair. As Bob Edwards of Citizens Energy Group in Indianapolis points out, potable water is the only utility that is necessary to live, which can be overlooked. Damage to a potable water pipeline can lead to contamination and numerous public safety and health risks.

Potable water may be safe to drink, but still contains dissolved gasses and solids. Over time, these naturally wear and tear on the pipeline and contribute to corrosion in metal pipelines, which is the same with hazardous liquid and gas.[22] This all makes material important to the utilities, where treated steel and plastics are favored and additional encasements are often used. It also underscores why external strikes like mechanical equipment from excavation can be so disruptive – it may hit at a weak point already worn down by time and corrosion. This may lead to a slow leak or a later rupture even years down the line.

Finally, storm systems and waste water infrastructure are often found quite deep. These are permanent fixtures intended to be kept well away from everyday life – and often carrying material that isn't volatile per se but is dangerous and can contaminate the environment and make people sick. Buildup of methane and other gases, however, can still lead to certain explosion risks. Concrete and steel often form these lines, sometimes found deeper than 10 feet under the ground.[23]

This small vantage point into the world below our feet is only that – small. But the reality is that there is so much infrastructure, carrying so many things, at such high volumes, that it is difficult to explain. So many individuals, companies, and government bodies work to keep it all functioning, as well as running them as business services.

In 1977, a U.S. DOT report[24] explained how even then the picture was complicated:

> Density of underground facilities range from a cheek to jowl proximity in city subterranean crossings to the isolated right-of-way of a transcontinental pipeline … The depth of burial is from a few inches to approximately 12 feet. In some cases there are regulations that specify the minimum burial distance while in other cases the frostline is the determining factor.[25]

Nearly 50 years later, that picture has grown far more complicated. If the list bullet point list above and previous explanations don't hit quite strongly enough for more visual learners, here is a snapshot of what is underneath a single intersection in New York City:

An open street in New York City.

Many Hats

While utilities may be lumped together to describe all buried cables and pipelines when discussing the big picture of damage prevention, each type of utility is its own distinct stakeholder group. This is important, because each of these stakeholder groups within the utility segment has their own specific needs and risks. The Common Ground Alliance (CGA) recognizes the following segments of the utility industry as separate stakeholder groups[26]:

- Electric
- Gas Transmission
- Gas Distribution
- Oil
- Telecommunications
- Public Works

We can consider water, sanitary sewer, and storm drains as separate. In many cases, but not all, one municipality may be responsible for all of these pipelines, but not always. Public works is a unique stakeholder group, because it may include just about every type of utility. For our purposes we are isolating utility groups based on the pipeline and cable function.

First and foremost, a utility company *owns or operates infrastructure*. However, utilities frequently assume additional responsibilities and concurrent roles beyond this foundational purpose.

When maintaining their own assets or laying new lines, utilities act as *excavators*. When others are planning excavations where their facilities are present, the utility may send an in-house locator – or member of the company – to the site. At those times, the utility is a *locator*. The intersection of these roles requires careful attention to detail, following proper procedures, and employing best practices.

One may assume a utility could skip notifying 811. After all, if they own their own assets, have facility maps, and their own locator, why would they notify 811 only to receive a notice back from them about their own assets? But they very well may interact with other buried lines at the site or near their own. Failing to notify 811 may result in accidentally striking or severing another utility's assets.

While excavators are commonly held to be the cause of any and all damage, and this is essentially a truism as they are the ones digging, the identity of that excavator matters for nuance. As we've explained, anyone digging is an excavator, and with many utility companies doing their own work, they are the relevant excavator in many instances. Maintenance excavation carries its own unique risks, because the excavator is actively looking for buried utilities, rather than avoiding them. Because of this, interesting findings emerge. As far back as the 1970s, it was clear that work on utilities is most prone to damage, with an official 1977 DOT report declaring, "The utilities cause the most damage because they must repeatedly excavate for maintenance, service installation, and repair of malfunctions."[27]

Lest anyone worry that citing a document from the 1970s is intended to skew the data, here it is from CGA in its latest DIRT Report, published in 2025:

> The top 10 known root causes by work performed reveal clear patterns that demand sector-specific attention. **Utility work dominates damage incidents across nearly all root causes**, with water/sewer and telecom/cable TV work consistently ranking as highest contributors to the leading root causes. This pattern underscores the urgent need for targeted interventions in utility operations, particularly given that these damages often involve facility operators (or their contractors) damaging each other's infrastructure or their own facilities during maintenance and installation activities.
>
> (Bold emphasis theirs)[28]

As we have already noted, "contractors" or "excavators" take the overwhelming majority of the blame, but knowing *which* excavator that is can complicate that basic picture, even if it remains a truism that "excavators do cause all the damage." Utilities working as excavators and striking their own lines may also be less likely to report, making true estimates of overall excavation damage difficult to know.

What Happens When a Pipeline or Cable Is Damaged during Excavation?

The physical repair costs vary dramatically depending on what is hit. Repairs should never be done by anyone other than the facility owner. The cost to repair a simple phone line drop is minimal, but repairs costs for high-pressure 42″ gas pipeline or an entire 288-count fiber will be enormous. The repair bill can reach into the tens of millions of dollars alone, which may only be a fraction of all the related costs. In Chapter 13, we paint a more detailed picture of all the direct, indirect, and societal costs.

Dane Lobb, PG&E:

> The biggest cost isn't just repairs—it's loss of trust. When our blue trucks roll up, the public blames us, whether or not we caused it.

Bob Edwards, Citizens Energy Group:

> Track everything—cut permits, traffic control, rentals, cleanup, even porta-potties. They all hit your bottom dollar.

When there is damage somebody is going to pay. The liability could fall on the shoulders of any of the companies involved, or be split up among the parties. With this in mind, the root cause of the damage is investigated in all but the most minor cases. The owner of the damaged facility, the company causing the damage, and the locate company (when it's a contract locator) all normally conduct their own investigation.

When a damage occurs, the number one priority is safety. The second priority is to document everything in detail, making no assumptions. Great damage prevention professionals also know that documenting a site before work is done is critical, and greatly aids the process if a damage does occur, when the type of details to be documented and the stress of the situation differ significantly. There are links in the chapter appendix to learn more details about damage investigation procedures, but here are the basic things that should be done:

1 *Ensure Immediate Safety*
    - Stop all excavation work.
    - If gas, fuel, or hazardous liquid is released, call 911, 811, and the facility owner/operator immediately.
2 *Secure and Document the Site*
    - Take photos, videos, and sketches of the damage, surrounding markings, and any evidence of mismarking or unmarked facilities.
    - Follow the "Clock Method": Take photos from 12, 3, 6, and 9 o'clock positions around the damage, like a clock face. Include both close-ups and wide shots from each angle. This ensures the location and extent of damage is fully documented.
    - Photos and videos need to be taken to show where the damage occurred compared to the marks. An oversized ruler should be used to show distances in the photos.
    - Record time, date, jobsite location, crew and witness names and contact information, and equipment in use, as well as proof of an active 811 ticket and corresponding positive responses.
    - Collect witness and crew interviews immediately while memories are fresh.

    > If a company doesn't have photos of the locate marks on the ground, the chances of successfully defending a damage decrease many times over.
    >
    > (Wayne Jensen, insurance/risk advisor)

3 *Track All Associated Costs*
4 *Keep documentation retrievable for at least 3–7 years, as claims often appear long after the incident.*
5 *Submit reportable incidents to the CGA DIRT database to strengthen industry-wide data to improve future damage prevention.*

> Retrievability of information is critical. Claims can come 12–18 months later. If you can't retrieve the photos, you're in trouble.
>
> (Fred Lesage, insurance engineer)

Unexpected damage to a live utility can stop an entire worksite for days or weeks. This is to say nothing of the disruption of service for consumers dependent on whatever was in that line. During the early stage of the process after a damage, the documentation and repair processes are the most urgent, while investigations may continue for many months, delaying completion of work and accruing more costs on the parties involved. Somewhat less consequential are non-live utilities. These may include, abandoned, retired, out of service, new but no product flowing, or de-energized or depressurized lines for some type of upgrade or maintenance. These are incredibly common, and they still slow down work considerably, even if they do not precipitate an emergency situation or claims process.

What Is Really Underground

Not all of the millions of miles of buried infrastructure consists of newly installed fiber optic cables or shiny new pipelines. Many utility lines are decades old, shifted underground over time, have been abandoned, or were never accurately mapped.

As federal spending increases fiber optic deployments, new populations demand services for natural gas, water, and power, new technology unlocks new ways of providing service, and utilities opt to underground their assets, more and more infrastructure is being placed beneath the surface. *Virtually all of this infrastructure is going in. Almost none of it is coming out.* With the exception of replacements of lead pipe or easy to remove lines that may still have value like copper wire, there will seemingly only ever be *more* buried utilities.

On the surface, it may seem like abandoned lines would be harmless, but as it turns out they can cause very serious safety hazards. Some of the significant challenges and risks abandoned lines pose include the fact that true abandoned lines are often electrically isolated, so standard EM locating can't pick them up. Unknown/abandoned lines can also stall potholing and bores, trigger redesigns, and add major cost.

Another risk associated with abandoned lines is that they can be confused with live lines. If there is no paint or flags indicating the abandoned line exists, but there are markings identifying a live line, the excavator may find the unmarked abandoned line and believe it is the live line. Believing they have found the marked line they may dig more aggressively away from the line they found and hit the live line.

When an excavator comes across one, they have to treat the unmarked lines as though it is live and waste time trying to determine the owner so they can get a definitive answer on whether the line is live. During an Excavation Safety Alliance Town Hall, Kurt Youngs, President, Youngs

Excavating; Board, Indiana 811, said: "We assume everything is live ... Even if we're 90 percent sure it's abandoned, it doesn't matter."

Arizona is a leader in dealing with abandoned lines, per Sandy Holmes, Executive Director, Arizona 811:

> Our law has required abandoned lines to remain on installation records for 36 years ... utilities must notify excavators, often using an 'A' in a circle ... and the 811 center must run an 'unknown line' process with a two-hour response.

Sandy Holmes also noted that "We seem to be the only state with an abandoned line identification process ... our law has required abandoned lines to remain on maps since 1980."

During that same Town Hall, Chris Stovall, President and CEO, Texas811; Co-founder and CEO, Linescape, explained that "In Texas ... we've sent ~40,000 digital locates [for abandoned facilities], zero safety incidents."

Many states have no verification or escalation pathway once an unmarked facility is exposed. Determining the best system to deal with abandoned lines is a work in progress for the industry.

Gaining More Utility from Utilities

The scale of underground infrastructure in the United States is vast – and still accelerating. With 50 millions of miles of pipes, cables, and wires already below our feet, and many more being installed each year to meet growing demand for power, water, gas, and broadband, the subterranean environment is only becoming more crowded. Unlike other systems that cycle assets in and out, almost nothing is being removed. Every year adds more complexity, more density, and more risk.

> We're going to see more and more pipelines being built, not fewer and fewer, and so it is in all of our best interest to make sure we're doing this in the safest way possible.
>
> (Sam Minifie, Policy Advisor, American Petroleum Institute, *ESA Town Hall*)

That reality makes damage prevention both more challenging and more urgent. Abandoned lines, shifting depths, and overlapping rights-of-way complicate excavation and planning. At the same time, innovations in sensing, monitoring, and coordination are transforming how we think about utilities themselves.

One example is the utilization of existing fiber optic infrastructure. Fiber optic cables utilize long strands of ultrapure glass to send information signals from end to end, primarily for internet and data services. Because these cables often contain dozens if not hundreds of individual strands, unused individual strands (known as "dark fibers") can be tapped for their ability to sense what is around them. These unused strands can be used to monitor for vibrations, changes in temperature, and more, sending data back to companies to help protect their utility lines.[29] Algorithms can be calibrated to classify a range of different activity, so that utility owners can know the difference between excavation vibrations, road traffic, or a pipeline leak. With specialized technology, fiber lines can act as a continuous sensor – detecting disturbances, triggering alerts, and even protecting neighboring lines. What was once a passive conduit can serve as an active safeguard, called a fiber optic sensor. This is especially interesting as telecommunication facilities are the most often struck buried line.[30]

The 811 Centers like Texas811 have already begun piloting the use of distributed fiber optic sensing to provide for greater situational awareness by cross referencing fiber alerts with open tickets (see the Texas811 Guardian Program).[31] In short, they can determine if excavation has begun on an existing ticket or notify utility owners that there is excavation activity near their assets when no ticket was ever called in.

As Mark Boxer, a technical manager with Lightera, explained in a recent panel discussion with the Alliance for Innovation and Infrastructure,

> It's almost never the first backhoe strike that actually causes the break. And so if you have enough time between the time that the backhoe hits the ground, you could potentially mitigate. There's actually been a case, I don't know if it's been published or not, but we've heard where a relatively sized, decent sized deployment was able to mitigate about half of the outages based on just using [fiber optic] sensing to help.[32]

This dual nature – utilities as both the assets being protected and as the tools for protection – captures the future of damage prevention. With every new mile of buried infrastructure, the risks compound, but so do the opportunities to embed resilience and intelligence into the very systems we rely on. The question is no longer simply what is underground – it is how we manage, monitor, and protect it in real time. The choices made today will determine whether the growth of underground utilities becomes a liability or a foundation for a safer, smarter, and more resilient infrastructure network.

Chapter 4 Resources

The CGA's Tips for Facility Owners

ESA/ACTS Now Town Halls

Improving Damage Investigation article by Ron Peterson

The True Cost of Damages. A checklist to help you make sure you account for all the costs associated with a damage, not just the obvious repair costs.

PHSMA's detailed review of the effectiveness of programs for the prevention of damage to pipelines by outside forces.

Oil Pipeline History

Notes

1 Kerr, K. (2015, March 30). Survey finds that nearly half of homeowners who plan to dig this year will put themselves and others at risk by not calling 811 before starting. https://commongroundalliance.com/sites/default/files/press_release_pdfs/2015%20CGA%20Omnibus%20release%20-%20FINAL.pdf
2 The 20 million miles statistic is mentioned several times in the NTSB proceedings of the 1994 workshop.
3 United States Department of Transportation; Research and Special Programs Administration; Office of Pipeline Safety. (1999, August). Common ground study of one-call systems and Damage Prevention Best Practices. https://primis.phmsa.dot.gov/comm/publications/CommonGroundStudy090499.pdf
4 Common Ground Alliance (CGA). (2018, April 3). Survey reveals nearly 40 percent of homeowners who plan to dig this year will put themselves and others at risk by not calling 811 before starting. https://commongroundalliance.com/Publications-Media/Press-Releases/survey-reveals-nearly-40-percent-of-homeowners-who-plan-to-dig-this-year-will-put-themselves-and-others-at-risk-by-not-calling-811-before-starting-3
5 U.S. Census Bureau. (2021, April 26). Historical Population Change Data (1910–2020). https://www.census.gov/data/tables/time-series/dec/popchange-data-text.html
6 U.S. Department of Housing and Urban Development. (1995, February). U.S. Housing Market Conditions Fourth Quarter 1994. https://www.huduser.gov/periodicals/ushmc/hsgmkt4.pdf
7 Federal Reserve Bank of St. Louis. (2026, February 4). Table Data – Housing Inventory Estimate: Total Housing Units in the United States. https://fred.stlouisfed.org/data/ETOTALUSQ176N
8 U.S. Energy Information Administration (EIA). (2025, April 9). Electricity explained. https://www.eia.gov/energyexplained/electricity/use-of-electricity.php
9 Pipeline and Hazardous Materials Safety Administration (PHMSA). (2025). Annual Report Mileage for Gas Distribution Systems. https://www.phmsa.dot.gov/data-and-statistics/pipeline/annual-report-mileage-gas-distribution-systems

10 Mohney, D. (2024, March 28). Fiber expected to add 23.2 million U.S. homes passed by 2028. https://fiberbroadband.org/2024/03/28/fiber-expected-to-add-23-2-million-u-s-homes-passed-by-2028/
11 Dierker, B., Rogers, O., & Murray, J. (2026, March). Measuring the Underground: Advancing a new estimate for the total linear mileage of buried infrastructure in the United States. https://www.aii.org/wp-content/uploads/2026/03/Measuring-the-Underground.pdf
12 Accenture. (2022). Best of utilities: summer reading list. https://www.accenture.com/content/dam/accenture/final/a-com-migration/pdf/pdf-159/accenture-utilities-summer-best-of-2021.pdf
13 Menduno, M. (2001, April 1). Plumbers Crack the Last Mile. https://www.wired.com/2001/04/plumbers-crack-the-last-mile/
14 North American Society for Trenchless Technology. (2023, March). HDD Good Practices. https://nastt.org/wp-content/uploads/2023/03/HDD-Course-Materials-2023.pdf
15 U.S. Department of Energy, Office of Electricity. (2024, July 23). Grid Communications Technologies. https://www.energy.gov/sites/default/files/2024-07/Grid%20Communication%20Technology_FINAL_optimizedv3.pdf
16 U.S. Department of Energy, Grid Deployment Office. (2024, September). Undergrounding transmission and distribution lines. https://www.energy.gov/sites/default/files/2024-11/111524_Undergrounding_Transmission_and_Distribution_Lines.pdf
17 Pipeline Safety Trust. (2019, September). Pipeline Basics & specifics about Natural Gas Pipelines. https://pstrust.org/wp-content/uploads/2015/09/2015-PST-Briefing-Paper-02-NatGasBasics.pdf
18 American Gas Association (AGA). (2025). A Natural Gas Primer: The Components of the Natural Gas Delivery System. https://www.mass.gov/doc/appdepdf/download
19 Environmental Protection Agency (EPA). (2024a, October 8). EPA Issues Final Rule Requiring Replacement of Lead Pipes Within 10 Years, Announces over $40.5M in Funding to Missouri to Provide Clean Water to Schools and Homes. https://www.epa.gov/newsreleases/epa-issues-final-rule-requiring-replacement-lead-pipes-within-10-years-announces-1
20 Plastics Pipe Institute. (2025). Water Service & Building Supply Lines. https://www.plasticpipe.org/Drainage/BuildingConstruction/-Applications-/Water-Service-Lines.aspx
21 International Code Council. (2021). International Residential Code for One- and Two-Family Dwellings § P2603.5, Freezing. https://codes.iccsafe.org/s/IRC2021P3/part-vii-plumbing/IRC2021P3-Pt07-Ch26-SecP2603.5
22 United States Geological Survey. (2019, March 1). Chloride, Salinity, and Dissolved Solids. https://www.usgs.gov/mission-areas/water-resources/science/chloride-salinity-and-dissolved-solids
23 Gieseke, M. (2020, February 25). Minnesota Department of Transportation Engineering Services Division. https://dot.state.mn.us/bridge/pdf/hydraulics/drainagemanual/tm2005b01-7061627-v2.pdf
24 Pipeline and Hazardous Materials Safety Administration (PHMSA). (1977). Effectiveness of programs for prevention of damage to pipelines by outside forces. https://primis.phmsa.dot.gov/rd/projects/15/
25 Pipeline and Hazardous Materials Safety Administration (PHMSA). (1977). Effectiveness of programs for prevention of damage to pipelines by outside forces (1 of 5). https://primis.phmsa.dot.gov/rd/FileGet/78/Prevention%20of%20Damage%20to%20Pipelines%20(1%20of%205)%2001-42.pdf
26 CGA – Common Ground Alliance. (2024). What we do. https://commongroundalliance.com/Membership-Engagement/What-We-Do
27 Pipeline and Hazardous Materials Safety Administration (PHMSA). (1977). Effectiveness of programs for prevention of damage to pipelines by outside forces (1 of 5). https://primis.phmsa.dot.gov/rd/FileGet/78/Prevention%20of%20Damage%20to%20Pipelines%20(1%20of%205)%2001-42.pdf
28 Common Ground Alliance (CGA). (2025). 2024 DIRT report: Analysis & recommendations. https://dirt.commongroundalliance.com/Portals/9/Common-Ground-Alliance-DIRT-Report-2024.pdf
29 Rogers, O. (2025, August 28). Enhancing pipeline safety and efficiency with distributed fiber optic sensing. Alliance for Innovation and Infrastructure. https://www.aii.org/enhancing-pipeline-safety-and-efficiency-with-distributed-fiber-optic-sensing/
30 Common Ground Alliance (CGA). (2025). 2024 DIRT report: Analysis & recommendations. https://dirt.commongroundalliance.com/Portals/9/Common-Ground-Alliance-DIRT-Report-2024.pdf
31 Texas811. (2023). Texas811 Guardian program. https://texas811.org/guardian/
32 Alliance for Innovation and Infrastructure. (2025, August 20). Aii expert panel | Challenges and opportunities in the data-energy triangle [Video]. https://youtu.be/tIwfDGMUw5c?si=5turwjA-Y-YgiK7m

Chapter 5

SUE and Mapping

"Get off of that darn tablet!"

This was almost a daily, if not hourly, complaint that Joe made of his nephew, Theo, when he was over for the weekend. He always seemed to have his head buried in a Game Boy or Switch or whatever the latest gadget was – he really couldn't keep up, but he heard the terms all the time.

Joe decided this weekend, he would put his foot down!

"Theo, we are going fishing!"

The groan that followed would have fit in nicely in any cattle pin in America. And the smile it elicited in Joe was one for the record books. The two went to the garage and grabbed a tackle box and walked down the back yard trail to the nearby lake.

Joe had a small flat-bottom aluminum bass boat with a trolling motor that he dipped into the water, once they had pushed off the sand and were past the first array of algae that had too many times wrapped itself all around the motor blades. Theo was starting to straddle the line of being mildly entertained by the new activity and being incredibly bored that the two would simply be sitting together in silence – or worse – talking!

Just before his breaking point, but not before one more prepubescent sigh, Theo caught a glimpse of something incredible: a screen! This was by design, as Joe wanted to show him something neat.

Fish!

"This is a fish finder," Joe said. "It uses sonar to show you where the fish are." The introduction of the technology had changed fishing to Theo, now it was a game. After learning how to operate the sonar, he would call out excitedly each time he spotted a group of fish. Joe smiled at his nephew's joy, feeling quite relaxed himself. Most of his job was underground, and the underwater space was a fun escape from work, while still resonating with what he liked about it.

The next day, Theo and Uncle Joe went downtown for a few errands. Theo out of habit pulled out his phone and was fixated on something. In a good mood today and not as troglodytic, Joe didn't say anything at first. But when Theo sprinted down the sidewalk and around a corner without saying a word, Joe was concerned.

"Theo! What are you doing?"

Theo was only two or three steps around the corner, and excitedly shouted, "A *Shiny Gible*!"

This, of course, was Greek to Joe. Apparently, it was a Pokemon or something. Joe asked Theo about it, and he told him that he can find and collect Pokemon all around the real world through his game. This gave Joe an idea.

DOI: 10.1201/9781003796619-6

For the rest of the afternoon, Joe used the app to show Theo things around town he almost certainly had never seen or thought of. As an engineer, Joe loved systems and technology that made life easier, and the ability to map out systems with technology was just about the best case scenario. For every *Rattata* he found, Theo also got a lesson in the nearby infrastructure landmarks. Sewer lines, manhole covers, street lamps, fountains, gas lanterns, cell towers – they mapped out the whole city with Pokemon Go.

No matter where they went, there was always more (and *more complicated* and *intricate*) infrastructure systems beneath their feet. Joe explained this all to Theo, but the only parts that stuck were the visuals. It was enough of a lesson for now – Theo was only 10 after all!

As an attorney, *sue* means something entirely different to me (Benjamin). To my dad, it'll always be a boy Johnny Cash sang about. But to anyone in the utility damage prevention profession, SUE can only mean one thing: Subsurface Utility Engineering.

That's a mouthful, which explains the succinct acronym, but it also sounds daunting. And perhaps the name does perfect justice to how daunting it really is. As the New York City picture from the last chapter illustrated, planning out, designing, building, repairing, and maintaining subsurface infrastructure can be a Herculean task.

SUE

SUE is the process of locating and mapping all the known and unknown underground utilities as precisely as possible in the design phase of a project – well before construction begins.[1] It also involves communicating and using that data to avoid or minimize utility conflicts with the project design (that's where some of the engineering comes in, otherwise its Subsurface Utility Mapping). SUE is very different from a locate requested through 811 Centers, which only provide an approximate location of buried facilities. SUE is more like giving the proposed construction site an MRI, like doctors do prior to surgery.

Jim Anspach, who is credited as the "Godfather of SUE," explains the difference between SUE and an 811 Center locate job with a very simple clear analogy: "One Call is the EMT rushing out to keep the patient alive. SUE is the preventive medicine that avoids the emergency in the first place."

If the reader will pardon one more analogy, the standard locating process involves a locator going to the site to mark what existing maps indicate is there, so that construction can begin in a few days. SUE, by contrast, is like a detective going out to the site well ahead of construction to explore the subsurface, use additional specialized technology, and actually create, bolster, or correct existing maps and documentation. They may create a Utility Investigation Report (UIR) or similar document.

Since the early 2000s, many people in the industry have been pushing for the widespread adoption of SUE. While not needed for every excavation, SUE is designed to reduce the risk of accidents and utility strikes well ahead of a project beginning. The Federal Highway Administration (FHWA) has encouraged the use of SUE since the 1990s, and the process is partially managed by the highly respected American Society of Civil Engineers (ASCE).[2,3]

A project utilizing SUE may be classified using a tiered system of quality levels – related to scientific uncertainty:

- *Quality Level D*: Information shown from existing records and information only, such as utility plans or GIS data.

- *Quality Level C*: Information shown on plans combines records with visible surface features, like manholes or valve covers.
- *Quality Level B*: Information on plans is based on field detection using sensing equipment such as GPR and electromagnetic locating, combined with existing records, to identify utilities both shown and not shown on plans.
- *Quality Level A*: Highest-level of accuracy, combines records, geophysical locating methods, and nondestructive exposure of utilities using potholing or vacuum excavation, among others.

For the damage prevention industry, SUE provides a proactive approach to utility locating beyond markings and can deliver higher accuracy that reduces the chances of incidents. SUE is not a piece of technology or a specific technique, but an overall engineering process.

> To put it kind of in layman's terms, essentially what we're doing is giving that information to designers and putting that cover or the photo on the front of the box of the puzzle, that puzzle being what's underground. Once you have that cover and an accurate photo of what's under there, not only is it valuable for the engineers [and] designers, which is where SUE kind of starts with, but also that puzzle can then be used for other aspects through construction and all that. And that's where I think that SUE component really fits in well as one of the first steps in damage prevention.
>
> (Lawrence Arcand, PE, P. ENG, 4Sight Utility Engineers, *ESA Town Hall*)

In addition to reducing the risk of a utility strike, SUE reduces the total cost of the project. A 1999 Purdue University/FHWA study showed that for every $1 spent on SUE, there was a $4.62 savings.[4] This translates into an estimated $2 billion in annual savings in highway construction in the United States. More recent studies have validated the relationship and demonstrated the strong return on investment. Specific dollar returns likely vary over time and by region, with numerous studies revealing positive returns, including Toronto in 2005 finding $3.41 saved for every dollar spent on SUE upfront, Pennsylvania in 2012 demonstrating $11.39 saved, and Louisiana in 2021 finding a $2.73 saved for every one dollar spent.[5,6,7] Along with improving public safety, SUE also provides benefits to society and everyone involved in these projects by reducing downtime, delays, and outages.

As Yogev Shifman, a co-founder of Exodigo explained in an interview, "Avoiding risk is cheaper than avoiding damage." Hoping to move the industry toward design requests and not days-before-digging locates, Yogev emphasized the importance of not being "reactive to the project, but proactive to form the project."

One hurdle SUE faces is that project developers must weigh the upfront design-phase cost against the temptation to defer until construction – essentially deciding whether to invest in SUE or "roll the dice" and address problems later. As Sam Wiffen, Founder and CEO of Reveal, explained,

> A mature developer would welcome the investment in SUE at the design phase, as they appreciate the ROI – risk, financial and safety. The real hurdle is the weighing up: do I spend money on this SUE thing, or roll the dice and see how we go when we break ground?

If they do not use SUE, the controlled investment is avoided during design, but new costs, risks, and potential impacts shift to construction – often in the form of redesigns, additional site work, delays, safety hazards, and other variable and unpredictable expenses. It is the age-old adage of, "pay me now or pay me a lot more later."

In reality, when there is no SUE work done in advance, contractors often raise their bids to cover changes they anticipate will occur. If they don't preemptively set their bids high, they may recover costs through change orders that ultimately exceed what SUE would have cost. Not doing SUE can also lead to project delays when unexpected pipelines and cables show up, or when they are in different locations than planned. The specific approach may vary depending on the contracting model – under lowest-price, build-only contracts, contractors typically push unexpected risks back onto the developer, while in design-and-build models, project delivery teams bear more of the risk and therefore have a stronger incentive to value SUE.

As Yogev elaborated, "We don't just show where utilities *are* but where they *are not.* And that area is priceless." Knowing there are no live utilities or abandoned lines will make project design and planning safer and more efficient, while guiding the more efficient planning of new infrastructure capacity.

Bridging SUE and Mapping

From this point forward, we turn from SUE itself to the broader world of mapping practices that build upon, but are distinct from, SUE outputs. Although often discussed together, the SUE process and utility mapping serve different but complementary roles. SUE provides a systematic, engineering-based approach to uncovering what lies below the surface and, in doing so, produces accurate and verifiable maps of underground utilities. These maps can be used effectively by designers to avoid potential utility conflicts in the first place, thereby avoiding the chance of construction damage.

Later in the project, these maps then become essential tools for excavators, locators, and project planners, helping to lower costs, reduce uncertainty, and improve safety. The challenge, however, is that many existing utility maps are incomplete, outdated, or inconsistent – highlighting why SUE is so valuable in creating reliable records and why continued improvements in mapping practices, especially through digital, precise, and accessible formats, are critical to strengthening the Facility Notification Center process and overall damage prevention efforts.

SUE professionals rely on maps to start their work, but ultimately produce better maps as their work product. So, if every buried facility were SUE-mapped, one may assume we have a permanent and perfect map, but that may not be perfectly attainable, because over time, utilities may shift, be manipulated, and new ones constantly added. This creates a continuous need for high-quality maps as new infrastructure is built, and a constant need for verified-SUE-based maps before major new work is undertaken. The industry must strive toward a cumulative documentation process with basic maps to start, SUE maps to base new projects on, then new infrastructure laid added into the high-quality SUE map that was used to perform the job. Digitizing this in the form of a "digital twin" is the ultimate goal of many in the industry, which may be owned or hosted by Facility Notification Centers, private vendors, or even government bodies.

The Challenge with Mapping

Unlike SUE, which delivers highly reliable, project-specific utility data, the mapping practices that follow – such as recordkeeping, GIS, and Facility Notification Center polygons – are designed primarily for communication and notification across broader audiences. They serve a different function but can be strengthened when informed by SUE outputs. For most excavations – like

SUE practice: Data gathered under direction of engineer is evaluated, synthesized, and overlaid with preliminary design to avoid or minimize potential cost and schedule impacts on projects.

your backyard project or a rural ditch – SUE isn't called for or utilized, but basic map records are consulted and used by utilities and locators.

Accurate map systems are one of the most important assets in damage prevention. After all, creating a process for facilitating the *locating* underground infrastructure is the ultimate point of Facility Notification Centers. While locators may utilize tools and advanced technologies like Ground Penetrating Radar to determine precise placement and depth of subsurface utilities, the information that gets them there comes from maps.

Even if maps of underground utilities exist, the accuracy and completeness of these maps are important. Despite the efforts of utility owners, locators, and other experts, the precise locations of many buried infrastructure components remain unknown. In particular, abandoned and old infrastructure is often unmapped.

Before computers and GIS, utility owners tracked their facilities using paper maps. In addition to potential degradation, these may employ difficult to read scales or terminology that is no longer in use or inconsistent with current records. Decades later, it's unsurprising that many of these records are lost or inaccurate.

> Records are a challenge and the bane of every owner operator. Mostly for PG&E, it is when we acquire someone else's assets. Suddenly we are responsible for what was in the ground and poorly mapped. I think of a military installation we acquired that had no

records. It really produced a challenge for us to go out there and update their facilities at our cost.

(Dane Lobb, Supervisor, Damage Prevention Public Awareness, PG&E, *ESA Town Hall*)

Yet even within a single company, record generation and keeping can be complicated.

The person in the field versus the person drawing the map may not be the same individual, and it can kind of become the telephone game, and the quality of the data there can suffer.

(Adam Zeciri, Locating Dynamics, *ESA Town Hall*)

It's easy to forget how easy access to maps has changed our society. Nowadays, we can type an address into our phone and receive directions and a time estimate for the journey in seconds. When was the last time you used a paper map on a road trip? Digital maps have had an even bigger impact on damage prevention.

[Before GIS] Any type of CAD drawing was done physically on big drawing boards. And then they'd roll those up and probably have a copy of them done somewhere and they stick them in a truck and then the guy would go out with those big drawings and go and locate.

(Duane Rodgers, CEO of PelicanCorp)

However, even with digital maps today, there is significant room for improvement. Typically, utility operators only share their detailed maps of underground infrastructure with trusted locators. Concerns over public access to these maps are twofold. First, there is a concern that public access to underground utility maps could lead to intentional damages from bad actors, whether it be terrorists, criminal organizations, or otherwise.

Second, there is also a belief that excavator access to utility maps will stop some of them from participating in the Facility Notification Center process. If an excavator has a map showing them where all the utilities are, they may think going through the whole process is a waste of valuable time. However, the fact that an excavator has access to a map does not make them a professional locator, raising the risk of accidental damages from misinterpreting maps or ignoring potential dangers. Not to mention that there may be other utilities present on a site that are unmapped or in a different database.

I think the fear for the longest time was that by sharing our [utility operator] data, you would encourage or enable people to not get locates, that they would just use the data and go. And I think what we've identified is there's always going to be that segment of excavators that aren't following the rules.

(Iain Stables, Damage Prevention Manager, ATCO Gas. *ESA Town Hall*)

However, there may be a valuable interplay between utility owners, locators, and excavators that is overlooked. In effect, SUE is a merging of all three of these disciplines, as it involves checking existing maps to locate buried utilities and verify with certain excavation processes – although SUE itself remains an engineering methodology. That may mean that working with existing stakeholder groups, particularly contractors, can effectively extract more value and help reduce more damage.

If you think about the people who are out there day in and day out. They're the ones that die. They're the ones that get hurt. They're a canary in the mine somewhat. And I feel like a very poetic solution to that would be to enable the canaries to solve the problem.

(Sam Wiffen, Founder and CEO of Reveal)

Unlike the literal canary, which may die in fulfillment of its purpose in the coal mine, the excavator does not have to perish. In fact, while already doing work and exposing lines, they could serve as a powerful resource in updating records. Applications like that used by RodRadar discussed in a previous chapter similarly provide new information that can bolster documentation during the regular course of excavation work. More collaboration is called for to achieve this value. As Yogev added, "what the industry is doing is important, but it is missing some of the right incentives."

Sam Wiffen resonates with this and goes on to explain that it may be longtime industry thinking that has boxed itself into a problem, when simply viewing things differently could resolve major pain points.

> So many times, construction crews are out there and they are the end of a very long trail of decisions that have been made or consequences that are in place and are the unfortunate recipients of the Swiss-cheese approach to risk. But if we can flip the opportunity where they are now in control of their destiny ... If we can create a value proposition that would enable them to solve the problem, whilst delivering their own work, that would be the fundamental shift in the way that we do this.

This multidisciplinary approach is resonant with other evolving trends in the industry. Not only can working across stakeholder groups produce more value, but it increases the skillsets and knowledge of each actor within a stakeholder group. This professionalization pushes industry participants into achieving mastery of more than just their own small slice of the industry. Utility-literate excavators are a desirable business to work with. Infusing more SUE training and opportunities into existing locator and excavator trainings, career advancements, and certifications can only benefit everyone in the long run.

In Chapter 11, you will see that this idea of cross-training is taking shape. Learning the basics of all the slices of the industry will make all stakeholders improve communication among stakeholders.

With new companies coming into the SUE space, more data will be available that will continue to improve maps. This can then be turned into additional valuable tools and resources. These new companies and their data may not be trusted by all stakeholder right away, and there is a balance needed to ensure the industry can continue advancing without leaving valuable improvements sidelined due only to their novelty.

Industry insiders have heard owner/operators not wanting to accept mapping/GPR/SUE information from outside parties due to liability issues. That makes doing the work and sharing the results a critical process that must be handled well. Chris Ross explains one solution to this very legitimate challenge:

> The Reveal solution is quite elegant as it retains the progressive updates and the history. The asset owner then has the ability to accept the updates depending on the level of trust. The thinking is that there is greater liability risk from retaining inaccurate records than a robust process to assess and consume updates.

While Reveal is operating in a traditional SUE role, others are working the problem from different angles toward the same goal. The non-invasive work of Exodigo leverages multiple sensors and sophisticated technology along with artificial intelligence, which can rapidly improve and correct record keeping for both accuracy and precision.[8] From there, companies like 4M Analytics and CivilGrid compile and aggregate data from many sources to form AI-powered maps and layered datasets.[9] In certain instances, these can even beat traditional SUE, such as AI identifying satellite imagery of a trench from a decade ago and incorporating new information about an abandoned facility that could be missed on a large-scale (e.g., five-mile) investigation.

Time will tell how these and other platforms add to the industry dynamic. While these innovations enhance mapping and record-keeping, they are not SUE themselves. Instead, they complement SUE by improving the accuracy, accessibility, and layering of records. Distinguishing between SUE as an engineering process and mapping as a record/data function helps clarify how these tools fit into the broader ecosystem.

Innovations Enhancing SUE Outputs:	Exodigo, Reveal, RodRadar, etc.
Innovations in Mapping Platforms:	4M Analytics, CivilGrid, etc.

While these mapping platforms represent new private-sector innovations, Facility Notification Centers are also adapting their own mapping systems to better serve excavators and utility operators. Once a utility owner or infrastructure developer knows what exists, they can plan work more strategically. But this may also work its way both upstream and downstream within the ecosystem. The Facility Notification Centers currently serve as a host for general-level utility locations within their state. Every 811 notification is intended to notify all member utility companies with facilities in that area, but the merging of SUE and AI-powered mapping platforms may ultimately have application at the call centers, should they elect to develop their own applications in concert with their members. The level of detail hosted at the call center will never be 100 percent, nor should it, as broad notice is the safest way to ensure notifications hit the right people. But some infusing of SUE-quality details and AI could strike a balance that reduces overnotification and improves efficiency.

This could then tie contractors back in by allowing electronic white-lining – overlapped with more accurate utility records on the call center's back end – to ensure all relevant utilities (but only the relevant ones) are notified for every dig. It is important to note that these operational maps are not the same as SUE-quality maps. SUE maps are created for project-specific certainty and design accuracy, while operational maps at Facility Notification Centers are meant to broadly inform and notify stakeholders of potential conflicts.

Importantly, data will only be able to be effectively shared if it is pulled together in real-time, or "on the fly" such as through FuzionView and limited to the dig site. A comprehensive nationwide map is not needed or desired by many stakeholders. Over-reliance on public resources, such as through FOIA requests or other records could result in outdated or stale records.

811 centers are already incorporating these innovative approaches today. In 2020, a group of experts formed the Underground Utilities Mapping Project Team (UUMPT) in Minnesota.[10] This organization subsequently merged into the Emergency Preparedness Committee of the state of Minnesota. From the work of this group, Gopher State One Call (GSOC) is now unveiling a "Site X-Ray" which they describe as:

> … automatically requesting geospatial infrastructure data from each participating facility operator that has utilities in the excavation area and displays the buried infrastructure for all participating utilities at a dig site.[11]

Specifically developed "to aid in the design and excavation process" GSOC also emphasizes that, "Site X-Ray is intended to supplement the One Call ticket process, but does not replace it. State requirements to identify the area of excavation and physically locate all underground facilities remain in effect." This demonstrates the ways innovation can work in parallel with existing law, and how industry stakeholders are constantly pushing the industry forward.

More broadly, and taking a step back from SUE, the way maps are used varies by stakeholders. Sometimes very precise maps are needed – such as for project design – while other times they serve as general roadmaps. The SUE discussion up to this point has focused in on a very specific kind of map – high-quality, narrow, and comprehensive for project designers and engineers. But as the industry thinks about what records it has and how to use them, there is also an effort to make certain maps more user friendly and to allow contractors to interface with them directly.

Using Maps to Lower Costs

As heard by damage prevention stakeholders of every type, communication is the key to reducing damages and maintaining safety. Maps are the ultimate visual communication tool, enabling the sharing of positional information. Even if utility operators are hesitant to make their maps publicly accessible, there is room for improvement within the Facility Notification Center system.

In practice, better mapping and facility awareness can influence project design decisions long before construction begins. Two brief examples illustrate how understanding the density of existing underground infrastructure can directly reduce costs and lower the likelihood of utility damage.

Wayne Jensen, VP – Director of Safety at Stahl & Associates Insurance, describes two great examples of how extra effort in planning projects can reduce both the cost of a project and the likelihood of utilities being damaged.

I once had a client who is a directional drilling contractor. They were bidding on a long-haul fiber installation from Tampa to Ft. Myers Florida in a "design build" format but with a "hard – not to exceed number" for cost to install. I told them that the most important piece of data they needed to have was the relative density of buried facilities on each side of the designated highway route. Obviously, installing on the side of the road with the lowest density of facilities was key and crucial to lowering costs and reducing the likelihood of any damage to buried utilities. I asked a friend who worked for a large national locator company if they could help. It was a pretty simple process for the locator to look at their mapping software and simply visually compare sides of the road for density and point out which side of the road had the fewest facilities. This advance planning allowed the contractor to be more confident in the bid they submitted.

One other project I was directly involved with, a local county put out a project for bid for installation of a 16″ ductile iron waste water line. The side of the road they picked for the installation it was nearly impossible to work around the degree of congestion. We researched the density of facilities on the other side of the road and there were almost no facilities the entire length of the installation. My owner bid the job based on asking for a redesign after the bid was let. All the other bidders submitted bids that were off the charts – maybe twice of our bid. We won the bid. After the award, we said to the county "you know it is unrealistic to install the line as it was shown in the construction drawings. Look at the other side of the road, let's redesign to install the line there." They agreed and off we went. They were elated with getting a reasonable bid to do the work and we made a good profit too.

There is such a huge need for both public and private project owners to design projects based on density of existing facilities so that new installations can be built with the least exposure to damaging existing facilities.

These examples highlight how even relatively simple mapping insights – such as understanding facility density along a corridor – can shape design decisions that reduce both risk and cost.

With most utility companies being required by state law to be members of their state, regional, or provincial Facility Notification Center, they already have to provide data on their assets. Historically, this was done on a grid model, where utility companies accessed or submitted to the back end of a call center software platform to functionally check the geographic box for where their facilities were. More recently, there has been movement toward "polygons" which are shapes drawn into the platform.

These are intentionally blunt representations, not sharp and highly precise SUE-type maps, because their intended function is to provide a buffer zone. If an excavator is digging with a crawler or backhoe, it is better that the company with a pipeline anywhere within 50 feet of the project be informed. Still, there is a level of precision that is important to reduce over-notification and improve ticket screening to filter out projects that genuinely do not pose risks to known utilities in an area. With better-quality maps on the back end, the front end accessed by excavators can be more functional.

Several Facility Notification Centers have begun using updated map software for excavators, both experimentally and available to the public. Rather than an excavator merely calling 811 and describing the dig site and project details, the excavator can draw it onto the map themselves – or electronic white-lining. This eliminates a middleman, but also relies on the accuracy of the excavator's representations. Despite continued debate, many stakeholders see the enormous benefit and potential in using maps for locate requests. By giving the excavator more precision in defining where they need locates completed, the locator also saves time by not marking additional, unneeded areas.

> One of the first things we noticed when we went live with our new Map It Yourself system was a drop in notifications going out. And there was a lot of concern at the beginning ... But the more we looked into it, the more it was like, wow, [the excavators are] just really exact in where they map ... In the interest of safety, an agent who's guessing or unsure of what they're saying is going to draw a much larger polygon to represent the dig area to make sure they're including as many members as necessary to be safe. Whereas the excavator is not worried about what's underground here on my map. It's just they know exactly where they're digging.
>
> (Sher Kirk, Operations Director at Utility Safety Partners, *ESA Town Hall*)

Additional mapping software need not put additional technological burdens on Facility Notification Centers and utility operators, and the benefits are substantial. Also, with all of the technological gains in recent years, it is easier than ever to incorporate these tools into the process. Any costs of implementation can be offset by the increased efficiency and safety.

> Members are tightening up their data that they're providing to us as well, they're no longer registering their data by grid, but instead by shapefiles or polygons. That is reducing notifications and reducing costs as well.
>
> (Mike Sullivan, President, Utility Safety Partners, *ESA Town Hall*)

Maps are powerful, but they only serve their purpose when they are both accurate and usable. As GIS and GPS systems advance, they bring immense potential to damage prevention, but they also introduce complexity. Not everyone in the field is a mapping expert, and a sophisticated system that cannot be readily understood or applied will fail to deliver its value. Accessibility, clarity, and training are therefore just as important as the precision of the data itself. In this way, maps are more than static records – they are dynamic communication tools, bridging the knowledge of those who design and manage infrastructure with those who put shovels in the ground.

This underscores the central challenge – and opportunity – facing the industry. The technology exists to map more accurately, share more efficiently, and prevent more damage than ever before. But unless the information is trusted, accessible, and effectively communicated, its promise will be lost in translation. The future of damage prevention depends not only on adopting advanced tools like SUE and digital mapping, but also on cultivating the professional standards, collaboration, and training needed to ensure those tools are used to their fullest potential.

	SUE	*Mapping*
Purpose	Precision project design	Communication/notification
Timing	Pre-construction	During design and throughout project lifecycle
Accuracy	Verified, tiered quality levels	Variable, often broad/general
Primary Users	Engineers, designers	Operators, locators, excavators
Cost Model	Upfront investment with high ROI	Ongoing operational necessity

Better Records Prevent Damage

SUE and high-quality mapping are no longer optional – they are essential tools for reducing risk, controlling costs, and improving safety across the industry. Their adoption reflects the broader shift toward professionalization, where proactive planning and accurate records define success more than reaction and repair.

As technology continues to evolve, integrating these tools will shape the future of damage prevention, not just by preventing strikes and saving money, but by setting a higher standard for practice. This standard is what will attract the next generation of talent, equipping them with advanced tools, deeper knowledge, and a stronger commitment to building a safer, more efficient infrastructure landscape. SUE is where the ecosystem converges, and it is that convergence that will solidify the future of communication and collaboration the industry relies upon today.

Chapter 5 Resources

ASCE 38-22 utility engineering standard and new companion standard ASCE 75-22

Aii Interview: A Vision for the Future of Damage Prevention (with Sam Wiffen)

ESA/ACTS Now Town Halls

CGA Technology Reports

The Minnesota Underground Utilities Mapping Project Team (UUMPT) meeting recording archive

UUMPT archive of miscellaneous videos for products and ideas they reviewed enroute to deciding to build FuzionView and Field Data Collector.

Technology Provider Examples: These are not endorsements, nor is it a comprehensive list, merely suggested resources to check out and learn more.

- 4M Analytics
- Dig-Smart, LLC
- Exodigo
- Irth Solutions, LLC
- KorTerra
- PelicanCorp & Geolantis
- Reveal (New Zealand)
- Ubrint

Notes

1 Federal Highway Administration (FHWA). (2022, November 1). Subsurface Utility Engineering. https://www.fhwa.dot.gov/programadmin/sueindex.cfm
2 Federal Highway Administration (FHWA). (2017, June 27). SUE Brochure. https://www.fhwa.dot.gov/design/sue/suebrochure.cfm
3 Federal Highway Administration (FHWA). (2017, June 27). ASCE Standard. https://www.fhwa.dot.gov/programadmin/asce.cfm
4 Federal Highway Administration (FHWA). (1999, December). Purdue University Study. https://www.fhwa.dot.gov/programadmin/pus.cfm/
5 Osman, H., & Al-Diraby, T. (2005). Subsurface Utility Engineering in Ontario: Case Studies and Lessons Learned. https://www.researchgate.net/publication/265994318_SUBSURFACE_UTILITY_ENGINEERING_IN_ONTARIO_CASE_STUDIES_AND_LESSONS_LEARNED
6 Jung, Y. J. (2012). Evaluation of subsurface utility engineering for highway projects: Benefit–Cost Analysis. *Tunnelling and Underground Space Technology*, *27*(1), 111–122. https://doi.org/10.1016/j.tust.2011.08.002
7 Mutoni, A., Zeringue, K., & Wilmot, C. (2021, January). Cost and Time Benefits for Using Subsurface Utility Engineering in Louisiana. https://www.ltrc.lsu.edu/pdf/2021/FR_644.pdf
8 Exodigo. (2025). Exodigo technology: Multi-sensor underground mapping software. https://www.exodigo.com/technology
9 4M. (2025). 4M Products. https://www.4manalytics.com/product
10 Minnesota IT Services: Geospatial Information Office. (2026). Emergency Preparedness Committee. https://www.mngeo.state.mn.us/committee/emprep/index.html#
11 Gopher State One Call. (2025, August 27). Site X-ray. https://gopherstateonecall.org/site-x-ray/

Chapter 6

One-Call Centers and 811

A loud bang rang out from the street. Jimmy sat up abruptly in bed. He peered down at the clock on his nightstand, but his eyes were too blurry to make out the hands. He took a sip of water from the glass and turned the lamp on to better focus his vision. It was only a few minutes after 5 a.m. He thought he may as well get up, and with his wife somehow not affected by the noise, he scurried out of the room quietly.

Stretching and letting out a yawn, Jimmy put a pot to boil on the stove and pulled out the can of instant coffee. As he waited for the water, he looked out on the quiet street. He would have been annoyed about the early wake up, but it was just Frank next door whose car was notorious for backfiring. And this morning, Jimmy appreciated getting an early start. Seeing the paperboy whiz by, Jimmy stepped out onto his porch and grabbed the bundle before making his coffee and sitting down in the kitchen.

Now with his steaming mug at his side and his horn-rimmed glasses perfectly resting on his nose, he spread open the paper. His face didn't flinch as his eyes scanned across the fold.

New High School to Break Ground This Fall in Growing Suburban District
Senate Debates Civil Rights Bill as Tensions Rise in South
Gas Prices Climb to 24¢ as Summer Travel Season Begins

The final headline felt like a validation. Jimmy was determined that June 1957 would be a swell time for him and his family. He had a new project ready to start soon, and it would be top notch when it was done. The only remaining challenge was to sort out the logistics.

His project was a brand new state of the art service station. Months ago, he had gone down to see Larry at the credit union and secured the loans he needed. He was back and forth to city hall a dozen times for permits and filings this spring, and now he was on the threshold of breaking ground! Today's task – and the reason an early start was quite peachy – involved calling all the local companies. He knew he'd be ringing the gas company, the electric company, the phone company, city water, sewer, and he would check in with the city on anything else he needed.

When the time came later that morning, Jimmy pulled up to the table and picked up the receiver. Placing his finger in the hole for "0" and making an arc, he placed his call to the operator. Without much delay, a woman's voice came through, "Operator, number please?"

Jimmy politely replied, "Yes, ma'am, I need Southwest Gas and Fuel, please."

"One moment, sir, I'll place that call for you. There may be a short delay."

"Thank you kindly, ma'am."

On her end, the operator plugs a cord from her illuminated switchboard into another slot. As Jimmy waits, she is connecting to the gas company and when ready, plugs in a second cord to connect them.

DOI: 10.1201/9781003796619-7

Once connected with Southwest Gas and Fuel, Jimmy explains that he'll be breaking ground on his service station soon and needs to make sure he doesn't disrupt any lines they may have in the area. He gives the plot number, and they talk about the project and agree to an in-person meeting. The conversation is brief, but cordial and full of details.

Jimmy plans to check with the electric company next. A bit unsure of the company – as it's across town from his home – Jimmy rings the operator again.

"Can you tell me the name of the electric company over in the Madison Court division?"

A wealth of local knowledge, the young woman in this small but growing suburban town knew just the right information.

This time, already seeing many blinking lights and cords on her end, she was as polite and composed as always, but had to tell Jimmy that it could be a few moments. When she rang Madison Light and Power, the line was busy.

"I'm sorry, sir, their line is busy. Would you like me to try again in a few minutes?"

Jimmy asked for the direct line to spare the switch board ladies another chore and thanked her for her help.

He repeated this routine another three times, each time describing the project and arranging with a local company to make sure that when he and his team got to work, they would not be creating any damage. It was a full day of work just doing all this coordinating, but Jimmy was by the book and wanted to protect his community just as much as he wanted to serve it – and as he had served it back in the war.

Once he had all his meetings with the local companies, he would get his builder on site to make his dream a reality. For Jimmy, this was the time of his life – building something new, working with people, and integrating himself even more into the community. He hoped one day his own family – maybe even a son or grandson – would build something of their own too.

A digital alarm sounded off for the third time in 10 minutes. JV finally managed to land his finger on the tiny portion of the screen that actually stopped instead of snoozing the alarm. When he did, his blurred vision made out 7:39 on the phone.

He rolled out of bed and stood in front of his open refrigerator for a good 60 seconds before realizing nothing looked good enough yet. He threw a coffee pod into the machine and set a mug underneath. His day was simple: call granddad, plan out the new deck, and get over to his friend's bonfire that night.

He knew granddad would have been awake for hours, so he didn't hesitate to knock that chore out of the way this early in the morning. Coffee in hand, he slid the lock button open on his phone screen and selected the contact from the icon with his address library.

"Good morning, James, how are we doing today?" the scratchy but chipper voice came through.

"I'm doing good, Granddad, happy birthday!"

While JV was using his brand new recently released iPhone, his grandfather was answering through a landline – albeit with a wireless receiver. Although JV had told him how cool these new touch screen phones were, it would never be enough to convince his granddad that he needed one. He just wanted to know the local news and read his paper. All the talk of stocks and global news flooding a phone screen did not appeal to him at all. And truth be told, JV scrolled past all that anyway, but loved that he had access to it if he ever wanted it.

"What do you have in store today, young man."

"Only making some plans for a deck I'm going to build on the house and then seeing some friends over on the beach."

"Make sure you let the power and water companies know you'll be digging!" his granddad made sure to remind him.

"Won't be an issue, I just have to contact 811 and they'll do the rest, shouldn't take more than a few minutes!"

Underground utilities related to water and sewage are as old as civilization. The first sewers were built more than 5,000 years ago in Ancient Greece and Mesopotamia.[1] The complicated aqueducts and filtering systems from Roman times are still studied today for their ingenuity and longevity. An argument could be made that had it not been for the use of underground water and sewage networks, early civilizations would have failed.

The undergrounding of utilities in the United States began in the Colonial Era, as emerging communities and cities made use of underground water and sewage systems. By the early 19th century, early gas lines were supplying cities, fueling street lamps through small pipes, even including wood pipelines.[2] In the 1860s, the first pipelines were placed below ground, and early telephone and electric cables were also being buried before the turn of the century.[3]

The practice of undergrounding continued to grow. By 1920, over 115,000 miles of oil and gas facilities stretched across the nation, most of it underground.[4] By the 1960s, a concerted effort was underway to increase undergrounding for other utilities.[5] Overhead power and telephone lines were the standard way to build, but often required extensive tree trimming and maintenance and added to visual clutter. Political folklore has it that Lady Bird Johnson was responsible for spearheading this movement as a way to beautify the cities and landscapes.[6] Aesthetics alone, however, were not enough to create a revolutionary shift in private market behavior.

Around this same time, utilities became aware of new vulnerabilities to their above-ground facilities. Exposure to the elements, collisions, and even vandalism all threatened electric transmission and distribution lines. These threats, plainly visible, inspired a push to protect critical infrastructure by moving it underground. Yet visibility itself had been a kind of protection – poles and wires out in the open are easier to spot, easier to avoid, and easier to repair.

Burying lines reduced many surface-level risks, but it introduced an entirely different one: out of sight can mean out of mind. Taking telephone poles out of the landscape may have beautified cityscapes, but those same systems still exist – just below our feet. And once hidden, they become more vulnerable to a new kind of accident. Whether it's a backhoe at a jobsite or a homeowner planting a tree, anyone breaking ground becomes a potential threat. Without clear knowledge of what lies below, a simple dig can quickly turn into a dangerous disruption.

As time went on in the mid-20th century, various utilities began offering free services, where excavators could contact the utility owner or operator to notify them that they would be excavating an area, giving the utility owner a chance to visit the site and mark where they may have facilities underground. This was done out of self-preservation for the utility, not as an amenity to construction crews, as the utilities ultimately sought ways to protect their own assets. It meant anyone digging had to have the initiative to come to them.

Marker posts and signs played a leading role as especially important lines like gas and electric facilities were buried. These included contact information for the local utility company operating the line. Often, if someone was going to dig and if thought to call utilities, they would likely only call the Bell Company, gas distribution company, electric utility, the city, or maybe cable TV. These would be individual calls made to each, which meant common oversights and omissions. Further, without good pipeline markers, the average person likely would not have known about many pipelines outside of major production areas, despite their growing ubiquity under the surface.

Today, many signs installed decades ago have faded as they became somewhat obsolete with digital marketing and the simple 811 phone number, even while they maintain standard direct phone numbers. This is particularly true of companies that have gone out of business or merged and of older telecom facilities. Pipelines are the only utility under federal obligation to maintain prominent signage for public awareness and regulatory safety compliance.

Originally, each utility may have viewed the *you call us* model temporary solution as ideal – after all, there is only one of them and many potential excavators – so a call-in service seemed simple. But for the excavator, there may in fact be many utilities, not just one to contact. Another problem was that if a builder wanted to excavate a lot, they may have to research all the utilities that provide service in that area, seek them out, and notify them individually by phone or even by visiting them. This was not only very time-consuming and costly, but failing to do this could be costly to both the utility owner and the excavator.

Markers have evolved significantly over the years, with many having warning messages that are visible from any direction. To be effective markers must be maintained and warning messages must be visible.

Even before these kinds of services gained broader traction, some municipalities had already started laying the groundwork for something better. Decades earlier, a few local governments and utility providers recognized that buried infrastructure came with its own set of risks – and that coordination, not just individual caution, was going to be essential. These early efforts weren't yet standardized or widespread, but they showed that the problem wasn't new, and the solutions weren't entirely novel either. Looking back more than a century, the early features of today's system come into focus – faint outlines that once seemed insignificant but now clearly resemble the fully formed structure we recognize today.

Starting from the Beginning

There were a few early attempts to avoid damaging buried pipelines and cables by creating organized ways to share information. Utility Coordination Councils (UCC) started in an effort to reduce damages and get utilities, municipalities, and contractors on the same page before construction began. A few early examples of this are the Board of Underground Work of Public Utilities of Chicago (usually shortened to Board of Underground), which was formed in 1910.[7] While it only had "voluntary membership and without statutory powers," it was a step forward and a key development toward the aim was to avoid utility hits and avoid conflicts. Notices and preliminary plans were shared among the members, which included all the Chicago Public Works departments, Cook County Department of Highways, Chicago Transit Authority, Chicago Parks & Sanitary District, Commonwealth Edison, Illinois Bell, People Gas Light & Coke, and Western Union Telegram.

Another early formal effort to coordinate construction activities began in 1926 in California.[8] The Los Angeles Substructure Committee, sometimes called the Substructure Damage Control Committee (SDDC), was created due to an uptick in road projects and utility burying programs that began after World War I, which were causing a spike in "dig-ins" to gas and telephone plants, along with traffic disruptions.[9] The process required the filing of a notice of proposed excavation by the agency or contractor planning to open a street, with instruction to plan to lessen traffic interference.[10] With a long history of permitting in Los Angeles, this most likely required printed documentation sent out by messenger – a courier errand sometimes accomplished on foot or even bicycle – to each Committee member, with each member then having a fixed timeframe to mark conflicts and return it. Over time, the SDDC even developed subcommittees dedicated to specific disciplines like "subsurface traffic interference."[11]

Years went by with minimal change in the form or function of coordinating, although local efforts around the country were beginning. Around the country, other examples of utility coordination at the very local level continued to arise.[12] The continued growth in population and the economy kept forcing the issue of dig-ins and excavation damage, such that even if solutions were put in place, more was always needed. Data on utility strikes and damages is difficult to capture even today, and historical data is virtually impossible to access, but the unfolding of history and actions by stakeholders confirm it was a growing challenge. Finally, by the 1960s, a system familiar to today's readers took its shape.

The First One-Call

From early days at Monroe High School in Rochester, NY, one may never have expected an unsuspecting stamp club member would go on to revolutionize the utility industry for a century. By earning a bachelor's degree in mechanical engineering from Purdue in 1943, this set the stage for Joseph "Joe" Kleinberg's to enter the utility industry and begin an over 40-year career that everyone reading this book depends on, even if his name is largely forgotten to history.

Several years into the job, Kleinberg was charged with restoring gas and electric service after several fires and explosions. That was in 1951, and while major incidents were still rare, this series led to multiple fatalities and 44 homes being destroyed.[13] This put damage prevention on his radar. A decade later, he was asked by his superiors to come up with a solution to get ahead of the damages.

In 1963, as a superintendent at the Rochester Gas & Electric Corp. (RG&E), Kleinberg and his colleague fielded nearly 5,000 local calls about excavation work in their area. Even with that call volume, 350 known incidents took place that year – which is essentially a utility strike every day – havoc for a local company. From these experiences, Kleinberg knew more was needed. He initiated the "Mac Mole" program to broaden the impact of excavator notices by providing a single-call option on all underground utility lines in the area.[14]

In 1964, Kleinberg set up that phone line to coordinate among multiple utility companies, including Rochester Gas & Electric, Rochester Telephone, and the Monroe County Water Authority. He served as chairman and had the line ring to RG&E rather than a new third-party UCC. The results spoke for themselves, and within 10 years, calls increased to 12,000, while incidents fell to 260 in 1972.[15]

His vision brought together local utilities, municipalities, and contractors into a cooperative communication system. Excavators called RG&E, whose representatives used teletype to notify member utilities. This would become the first true One-Call center.[16] In addition to establishing the first functional process, this also launched the first marketing mascot concept. In Rochester, NY, it was "Mac Mole" and later, Detroit, MI would create "Gus Gopher." These digging-centric names reinforce the idea of excavation, while later concepts have expanded this further with names and personalities like Miss Dig, JULIE, and even educational characters like "Hazardous Matt."

The single-call concept reduced accidents and improved coordination dramatically. Now, excavators planning to dig on a lot could call a single number and be connected with a receptionist, who would let each utility know about the planned excavation. This system of calling one place to find the locations of multiple buried utilities became known as a One-Call Center. Captured by the Congressional Record of the U.S. Senate in 1973, Kleinberg optimistically predicted that soon, "15 or 20 such systems will be operating across the country."[17] His optimism was well-placed.

The concept of a unified phone line soon took off as an efficient and valuable practice. This huge innovation at the time solved a major problem. As it was refined over the next decades, more and more damages were avoided and infrastructure was safeguarded. Yet, before many more One-Call centers could pop up organically, national organizations began providing top-down guidance.

The result was a combination of organic local cooperative organizations and federal and state directives, ultimately forming the unique One-Call center dynamics we see today: mandatory organizations dictated by both state and federal policies, operating as local and regional industry-run nonprofits.

National and Federal Involvement

In 1965, there was an incident involving a 24-inch natural gas pipeline that ruptured near Natchitoches, Louisiana.[18] With 17 fatalities and at least nine additional injuries as a result of the incident – along with a 10-foot deep crater and five homes entirely leveled – it was described as the worst pipeline accident since the 1950s. While this was not excavation-related damage (it was a high-pressure corrosion failure), it spurred greater public awareness of pipeline hazards and laid the groundwork for later excavation-damage prevention laws. In fact, "dig-ins" and pipeline strikes were already a well-known issue and were already being discussed in concert with broader pipeline incidents and vulnerabilities. Stakeholders planned to address them together.

In fact, an Ohio Public Utility Commissioner, remarking on the nascent success of One-Call centers, believed that they would not be enough.

> A number of these pipes have been in the ground for 25 or 30 years, and some have been down there a lot longer than that. The action of the elements – soil, acidity, water, bacteria – are not really known in a lot of areas. The probability of a disaster is likely to increase with the passage of years.[19]

So with a major incident involving corrosion, and known local examples of excavation incidents, Congress acted to ensure this critical infrastructure was securely overseen by federal authority. In 1967, Congress established the National Transportation Safety Board (NTSB) as an independent agency with a mandate to investigate major transportation incidents involving aviation, highway, maritime, pipeline, and railroad, with particular emphasis on any involving the transportation of hazardous materials.[20] Accident investigations provide critical insight into what went wrong, while studies and recommendations would prove to be the earliest major contribution of the NTSB.

In 1968, President Lyndon B. Johnson signed into law the Natural Gas Pipeline Safety Act, giving the Secretary of Transportation the authority to regulate hazardous liquid pipelines.[21] This was in the early days of the department, as President Johnson only signed legislation consolidating programs into a new Department of Transportation a year earlier.

The delegation of authority was important but did not reach to other utilities or local construction practices. Still, it enabled a new focus on pipeline safety procedures that would go on to mature into damage prevention programs.[22] Unfortunately, across the country, incidents continued to affect pipelines.

In 1973, the NTSB released the results of a highly anticipated special study, *Prevention of Damage to Pipelines*. Motivation for the study had accelerated after several serious pipeline incidents in the late 1960s and early 1970s. The study concluded that the primary cause of damage to pipelines was excavation without adequate knowledge of the pipelines and their location.

> The primary cause of pipeline accidents is the damage from earth-moving and excavation equipment.
>
> (Prevention of Damages to Pipelines, 1973 NTSB Report)[23]

Another complementing report released by the USDOT was clear that pipelines are specifically at risk, but not uniquely at risk:

> The problem is not limited to gas pipelines, for other organizations operating buried waterlines, sewer lines, and electric and telephone cables have a high incidence of dig-ins.[24]

Based on these findings, the NTSB issued 17 critical recommendations to eight organizations.[25] While recommending legislative changes to mandate that pipeline operators establish damage prevention programs was among the most specific, others encouraged development of new voluntary standards and coordination. Three further recommendations would advance the pace of damage prevention for the public at incredible speed:

> **To the American Public Works Association:**
>
> Encourage its local chapters to establish Utility Coordinating Committees in all urban and suburban communities where effective committees are not currently in operation.[26]
>
> Coordinate … the establishment of a national organization of Utility Coordinating Committees.[27]
>
> **To the National Association of Regulatory Utility Commissioners:**
>
> Urge its member commissions to encourage the establishment of local and statewide Utility Coordination Committees where non exist [sic].[28]

These effectively urged the creation of notification systems (One-Call Systems) and safe excavation training, with additional recommendations that these organizations support further legislation that would mandate additional damage prevention practices. This boosted interest and the pace of the creation of One-Call Centers.

Elsewhere at a national level, in 1974, the APWA joined forces with the American Society of Civil Engineers (ASCE) to conduct a study, "Accommodation of Utility Plant Within the Rights-of-way of Urban Streets and Highways: State-of-the-Art" for the U.S. Department of Transportation, Federal Highway Administration (FHWA).[29] The study was extensive, including surveys of 500 agencies (with 222 responses), interviews with 40 communities, and input from utility associations, FHWA, and other national organizations. From a damage prevention perspective, the main impacts of the study were a push for legislation and a best practices guideline for improving coordination between local government agencies and utilities. This study made it clear that utility activities in the nation's rights-of-way needed to be coordinated. As a result of this study and others, the APWA officially formed the Utility Location and Coordination Council (ULCC), which was eventually renamed to the Utilities & Public Rights-of-Way (UPROW) Committee.[30]

With this study coming on the heels of the others from the NTSB, momentum for legislation and the creation of One-Call Centers was in full swing. In the early 1970s, One-Call Centers were launched in Michigan, Ohio, Pennsylvania, Arizona, Georgia, California, Maryland, Delaware, and New England. All this activity along with the creation of the then ULCC led to a 1975 meeting of 44 representatives from 22 states to share ideas and discuss ways to improve their fledgling systems. It quickly became clear that there was a huge upside to sharing information and ideas, so they decided to form an ongoing committee which became known as One Call Systems International (OCSI).[31]

OCSI created a logo, which included a simple, clear symbol warning people not to dig in the area. This symbol, without the OCSI name, became the common denominator on many pipeline and cable warning signs and decals. *Think you can guess it? Check your knowledge with the code.* OCSI trademarked the symbol and logo but allowed its free use for warning of underground utilities. This was a significant step toward standardization across the industry. Additionally, by using a symbol that could be understood without words, language barriers were bridged. OCSI is now a committee within CGA.

By 1976, over half of states had proposed or enacted damage prevention legislation, while nearly one-third had statewide One-Call systems in place, while the majority of states either had only local efforts or none at all.[32] However, the nation still lacked uniformity in marking practices. As marking the location of buried facilities started to become common, different utilities and municipalities used separate color code systems, or did not use one at all. This led to confusion, utility strikes, and delayed excavation work. While the notification centers do not do the marking, as the central stakeholder facilitating the process, having a standardized color code across the nation was undoubtedly needed to consider the one-call concept complete and universal.

All the while, there was an outstanding recommendation from the NTSB's 1973 report to "Develop standard colors for identifying underground facilities to be used for temporary marking and staking by operators of such facilities, and urge local chapters to support adoption and use of these standard colors."[33]

To solve this problem and finally fulfill the NTSB recommendation, the APWA officially adopted the Uniform Color Code for Temporary Marking in 1977.[34] This simple solution was a big step forward and was eventually incorporated into federal guidelines. With this in place, no matter which notification center or in which state an excavator made contact with, the same colors would be used by the local locator.

By the 1980s, the trend of only more localized coverage was giving way to statewide coverage as we see today, driven largely by new legislation. As each state passed their own One-Call laws, many of the provisions had similar components, yet were largely distinct from one another and continued to evolve.[35]

Many of the common requirements included 48-hour to three-business-day notice prior to beginning excavation, use of the APWA color code for temporary marking, some form of penalties for non-compliance, positive response (facility owners must let excavators know when they have marked their lines), and systems to handle emergency excavation needs.[36] Likewise, to be functional in the first place, many laws require underground facility owners to provide mapping or basic geographic information on their facilities to the One-Call center. Without this information, the One-Call center would not be able to inform facility owners about planned excavation near their lines. However, just because the One-Call center has this basic geographic information does not mean they are in charge of locating facilities, or even capable of it. The maps provided by facility owners to One-Call centers for notification purposes are typically grid maps, where a One-Call center can see if a specific operator has utilities within a grid square, but not what type, how many, or specific locations.

Within these varied laws, some states exempted specific types of digging in certain circumstances. Examples include agricultural tilling (within specific depths), homeowner hand digging, railroad maintenance, grave digging (for pre-designated plots), and emergency excavation.[37,38] Some laws also exempted certain entities from mandatory membership – meaning these entities do not have to participate in membership responsibilities like paying dues or providing maps to the One-Call. Commonly exempted organizations include State Departments of Transportation, Railroads, small towns, irrigation districts, and gas and oil gathering systems.[39,40]

While the 1970s and 1980s were active for damage prevention legislation, the pace slowed considerably as time went on. After 2000, major changes to dig laws have come in response to critical incidents or after major technological changes have been well established, like the proliferation of the internet enabling new communications practices.

The result of the buildup of damage prevention laws across the country over multiple decades and slowdown in revising and tightening them has also meant many exemptions remain in effect. Exactly which activities qualify as excavation remains an educational challenge today. The 2023 CGA DIRT Report links a disproportionate number of damages to activities or entities operating under legal exemptions.[41]

Every state is different, so one cannot assume these are the rules in one's own state. The following link takes you to the Excavation Safety Guide Resource Directory, which summarizes each state and province law and lists further state law details.

State laws are the purview of that state's own legislative bodies, yet federal rules have strongly shaped these state laws. One-call centers are required to exist in state laws on the basis of federal rules. While up to this point, many One-Call centers were already being established or required to come into existence, through pipeline authority, the USDOT helped close the gap to ensure the entire country was covered through state programs and to increase participation.

More Federal Action

In 1982, the federal government established requirements for pipeline companies to maintain a Damage Prevention Program through CFR 49 Part 192.614.[42] CFR stands for the Code of Federal Regulations, while title 49 designates the DOT as the enforcement agency. The 192 (also called "Part 192") refers to the Pipeline Safety Regulations for the Transportation of Natural and Other Gas by Pipeline (managed today by PHMSA under the USDOT). Finally, Section 614 is the specific rule section within Part 192 that requires Damage Prevention Programs for pipeline operators.

The reader may not be surprised to learn this was a somewhat belated fulfillment of the NTSB recommendations years earlier (from the same report as above), calling for

1 The Office of Pipeline Safety of the Department of Transportation:

  a Amend 49 CFR 192 and 49 CFR 195 to require each pipeline operator to establish a program for the prevention of excavation-type damage to its underground facilities.[43]

The recommendation is footnoted with "This recommendation is similar to Recommendation 1(a) in the Burlington, Iowa, pipeline accident report," which you could read as the NTSB saying "per my last email." It is very common for the NTSB to issue a recommendation that is directly related to safety and still takes years before it is acted upon. In this case, it was nine years.

Before CFR 192.614, damage prevention for pipelines was largely voluntary. This rule made it mandatory for pipeline operators to proactively engage in excavation safety. The rule also made One-Call participation mandatory for pipeline operators. While these rules, enforced today by PHMSA, could only hold pipeline operators responsible, they did provide guidelines that other facility owners could look to as a standard.

CFR Part 192.614 impacted One-Calls and damage prevention by formalizing damage prevention responsibilities, promoting One-Call participation, and introducing the first federal requirement for public education campaigns directed at excavators and the general public. This set the stage for many state One-Call laws even though it only had jurisdiction over pipelines.

For the next decade, the impact of these federal rules helped spur growth again from the bottom up. As the industry was growing, it became clear that sharing ideas was the best way to become efficient and provide the best service possible. In the early 1990s, the Southeastern One Call Systems (SOCS) was formed by the One-Calls in Florida, Georgia, Tennessee, South Carolina, North Carolina, Alabama, Arkansas, Louisiana, and Mississippi. The purpose was to bring One-Call Executives and One-Call Board members together to discuss industry trends and leverage opportunities in the Southeastern Region. Since its inception, Kentucky has been added to the group.

In 1986, One Calls of America (OCOA) was created by Michigan, Ohio, Alabama, Tennessee, and Colorado, with the goal of combining their telecommunications volume to leverage greater efficiency.[44] At the time, all participating states used AT&T long-distance service, and higher call volumes translated directly into larger savings. OCOA leveraged this model to expand into Health Savings, Retirement Savings, Equipment Purchase Savings, etc. Through these initiatives, OCOA helped lower the cost of operating a center without compromising service quality.

In 1998, the federal government infused the next jolt of energy into the damage prevention industry. It is hard to say which of all these actions has been most consequential, but this one was arguably one of the most recognizable to today's industry stakeholders.

Congress penned the Transportation Equity Act for the 21st century, which required the U.S. Department of Transportation's Office of Pipeline Safety (OPS, now PHMSA) to identify and evaluate existing damage prevention practices.[45] This led to the influential and well-known *Common Ground Study*. Over 160 stakeholders participated in the study including utilities, contractors, locators, regulators, and trade associations. In 1999, the final report, officially titled "Common Ground: Study of One-Call Systems and Damage Prevention Best Practices," identified 132 best-practices, including use of One-Call notification systems, accurate locating and marking standards, excavation techniques near underground facilities, public education and awareness, and enforcement of One-Call laws.[46] The report recommended forming a national damage prevention organization, which ultimately led to the formation of the Common Ground Alliance (CGA) in 2000.[47]

> At the heart of damage prevention is improved information accuracy and consistency in communication between excavators and operators of underground facilities. One-call systems provide a reliable and efficient process for excavators to notify facility owners/operators of planned excavations.
>
> (Common Ground Study, 1999)

The OPS provided the seed funding to create and launch the CGA. Since its creation, CGA has been funded primarily by a combination of private industry sponsorships, member organizations, and funding from PHMSA. The CGA played an important role in bringing the locator, operator, and excavator stakeholders together on a national basis.

Bob Kipp was hired to be the CGA's first President in 2001. He went on to lead CGA for almost 16 years. The CGA was created based on a push and funding by PHMSA, so Mr. Kipp's extensive experience in the telecommunications industry helped provide a more diverse perspective. Under Mr. Kipp's leadership, CGA created the industry's first Best Practice guide, the first nation infrastructure damage data tracking system (i.e., the DIRT platform), and the launch of 811 as a national number for requesting a locate. In recent years, the peerless Sarah K. Magruder Lyle has served as CGA's President and CEO, leading the organization to growth and financial success.

Many of the One-Call Centers had participated in a national organization, OCSI, since the late 70s, but the CGA brought all stakeholders together. The primary initial task of the CGA was to create Best Practices that every stakeholder group would agree to. These Best Practices are continually updated and are the industry standard.

> Through the implementation of these best practices, one-call centers can evaluate their operations to provide better communication between excavators and facility owners/operators. This will accomplish the goal of protecting the public, excavators, and the environment and preventing disruptions to public services and damages to underground facilities.
>
> (Common Ground Study 1999)[48]

Building on the foundation of new collaboration and shared standards, federal lawmakers soon took further steps to strengthen pipeline safety and damage prevention.

The Pipeline Safety Improvement Act was passed in 2002 after several high-profile deadly pipeline accidents in the late 1990s and early 2000s, including a 1999 gasoline pipeline rupture that killed three people and caused major environmental damage in Bellingham, Washington and a natural gas transmission pipeline explosion in Carlsbad, New Mexico in 2000 which killed 12 campers.[49] This law further expanded the role of pipeline companies and One-Call centers in providing education programs and public awareness about safe digging. There is ongoing realization of this requirement even today.

> We'll build an outreach program and reach out to these excavators and offer free damage prevention presentations. So we'll go out and educate these contractors, let them know what we can do to help them if they're having problems with locates or if they're having problems with facilities being in their way or whatever it is we can do to help them.
>
> (Jerry Cobenais, Operations Manager at Excel Energy, *ESA Town Hall*)

With greater One-Call participation and better excavator training and awareness, the next metric of effectiveness is actual damages. If that isn't trending downward, recalibration is needed.

The law also required the collection of damage data, which led to the Common Ground Alliance Damage Information Reporting Tool (DIRT) being launched in 2003.[50] The CGA DIRT program relies on voluntary anonymous submissions of damage reports to identify the root cause of the damage. The CGA DIRT program was modeled after the Colorado system created by a 2000 state law "Underground Facility Damage Reporting Specification," which requires the reporting of damages by all owners and operators of underground facilities.

The CGA analyzes the data annually and produces a report, which is the industry standard for tracking damages and looking for trends. While the voluntary basis of the damage reporting impacts the accuracy and completeness of the data, the DIRT report is the global standard for damage tracking. Without good data on what causes damages, it is difficult for the industry to take actions to reduce damages in the future, and DIRT provides this data. Included in the annual DIRT Report are incoming locate request volumes and outgoing utility transmissions, which the call centers provide, helping inform the industry overall on the state of system use and its relationship to damages. Not only does the annual DIRT Report regularly offer recommendations to One-Call centers, but those centers also use the information and learnings from the report to improve their procedures, training, and outreach.

By 2005, a comprehensive and cohesive damage prevention industry was finally operating on a national level. Not only did every state have a One-Call center, but those centers along with all of the key stakeholders joined together alongside the CGA. New data tracking and damage reports were helping hone practice and procedures at the call center level, but the final federal salvo was still to come. If the Common Ground study is well-known to industry players, the next move truly made the industry a household concept. It takes a step back in chronology to show how it unfolded, but it brings us to the present quickly.

Through years of advocacy and collaboration by CGA and other stakeholders, alongside congressional champions from across the country, the FCC designated 8-1-1 as the universal national number "Call Before You Dig" number and mandated that 811 be functioning in all exchanges in the United States by April 13, 2007, which was a very aggressive timeline that created some controversy and stress.

CGA has credited Congressman Chris John (D-LA), a member of the U.S. House Energy and Commerce Committee representing Louisiana's 7th District from 1997 to 2005 and president of the Louisiana Mid-Continent Oil and Gas Association, as a key champion. Rep. John had a deep focus on energy infrastructure and pipeline safety and likely worked behind the scenes on the effort.

In total, 43 U.S. Representatives co-sponsored the Pipeline Safety Improvement Act of 2002 in the House of Representatives and nine Senators sponsored legislation in the Senate.[51] Each of these champions recognized the confusion caused by multiple state-specific One-Call numbers, and saw a strong need for a standard, easy-to-remember number to increase public compliance and reduce utility strikes. The law "Provides for establishment of a three-digit nationwide toll-free telephone number to be used by State one-call notification systems." While it took three years for the FCC action and another two years to realize its implementation, this was virtual warp speed for the federal government, and the actions were monumental in practical reform for the industry.

811 completely changed the industry forever. Not only did national public awareness become eminently more feasible, but One-Call centers received a new phone number (although technically they still had and have specific local phone numbers). To some, the designation of 811 as a national phone number is what completed the concept of One-Call centers, serving as the keystone to cap off the effort. With all the final pieces coming into place, the industry had everything it needed to succeed – and the federal government wanted to ensure all these investments and rule changes were effective.

In 2006, in the course of reauthorizing PHMSA and bolstering its jurisdiction, Congress passed the Pipeline Inspection, Protection, Enforcement and Safety (PIPES) Act.[52] This gave PHMSA the authority to evaluate state One-Call/damage prevention programs for effectiveness and to take federal enforcement action in states with inadequate enforcement. From everything we heard at the time, PHMSA's goal was to do the evaluations and have any state not meeting their requirements make the necessary corrections without PHMSA needing to step in to force changes.

The key components in the PIPES Act were the "Nine Elements of an Effective State Damage Prevention Program" which created the standard that PHMSA would use to evaluate state programs.[53] Many states were already doing many of the components outlined, but these were intended to create a national minimum standard. The actions covered included education and training, participation in the One-Call system, communication, and enforcement. More details on these nine elements can be found later in this book and in the chapter resources.

On May 1, 2007, 811 was officially launched in a ceremony in Washington, D.C.[54] The Secretary of Transportation, Mary E. Peters attended the launch event to highlight the importance of this step toward improving public safety. Scott, the 811 task team, and a handful of other members of the CGA Education Committee, along with CGA President Bob Kipp attended the ceremony. With only a few remaining issues to solve, 811 was officially operational across the entire country.

The creation of 811 did not expand the availability of One-Call Centers; the same number of centers existed immediately before and after, but it made national public awareness campaigns possible. Until May 1, 2007, every One-Call Center had its own exclusive toll free phone number to call. If anyone wanted to run a national media campaign on damage prevention, they would need to list all the different phone numbers and One-Call Centers. The 811 charged all that.

Even the naming conventions for these centers began to shift in response to the nationwide number. The evolution unfolds further ahead and continues to today.

National education was also made ever easier by the CGA because they developed a logo for 811 and made it available to all stakeholders to use. In addition, the CGA created educational materials, ad templates, and PSAs which were also available to stakeholders. This was critical because this made consistent messaging easy. The 811 also made it much easier for contractors who worked in multiple states. The value of 811 at the time cannot be overstated.

In 2009, PHMSA initiated an effort to assess the extent to which each state was incorporating the nine elements of effective damage prevention programs. This was called the State Damage Prevention Program Characterization (SDPPC) initiative.[55] While the SDPPC initiative itself was conducted only three times, with its last update in 2014, PHSMA continues to track states without adequate damage prevention efforts through other reviews. As of September 3, 2024, only three states were deemed to be inadequate by PHMSA in terms of their pipeline enforcement programs.[56]

Notable Events in U.S. One-Call History

Like most industries, there were some pioneers that really turned the idea of a One-Call center into reality. These people championed new ideas and turned ideas into reality. Without these people the industry would not be what it is today. The next generation will undoubtedly transform the industry into something unimaginable at this point, but it is the foundation created innovators like Bill Kiger, Pennsylvania One Call; Tom Hoff, One Call Concepts; Michael McNamara, One Call Systems; Hasmukh Parikh, Northfield Data Products; Phil Johnson, Virginia One Call; John Shelton, Florida One Call; and Claudette Campbell, Georgia One Call to name a few.

- *1964*: First One-Call center launched in Monroe County, NY.[57]
- *1968*: Pipeline Safety Act signed into law by President Lyndon B. Johnson.[58]
- *1970*: MISS DIG was established in Michigan as the first statewide and 24/7 utility protection notification service in the United States.
- *1973*: NTSB commissioned a study recommending centralized damage prevention coordination.[59]
- *1974*: Accommodation of Utility Plant Within the Right-of-Way of Urban Streets and Highways study.[60]
- *1975*: 44 representatives from 22 states met in Pittsburgh to form what would become OCSI.[61]
- *1977*: PA One Call registers the first national One-Call logo as a registered service mark.
- *1977–2000*: Annual symposiums held, hosted in major cities, spreading best practices.
- *1977*: OCSI (No Dig) logo was adopted, with trademark filed in 1980 and registered in 1983.[62]
- *1980*: The Uniform Color Code for Temporary Marking was formally adopted by the American Public Works Association (APWA).[63]
- *1982*: CFR 49 192.614 codifies use of One Call systems for pipeline damage prevention.[64]
- *1985*: OCSI became part of the APWA.[65]
- *1994*: SOCS was formed.
- *1996*: OCOA founded.[66]
- *1999*: The Common Ground Study was published.[67]
- *2000*: The CGA was formed.[68]
- *2002*: Pipeline Safety Improvement Act was passed.[69]
- *2003*: CGA DIRT program was launched with the first report published in 2004.[70]
- *2005*: The FCC designated 8-1-1 as the universal national number "Call Before You Dig" number.[71]
- *2007*: 811 was implemented nationwide.[72]
- *2009*: Beginning of PHMSA's SDPPC initiative.[73]
- *2019*: OCOA was dissolved and the Facility Notification Center Association (FNCA) was founded in 2020.[74]

Present Day and Operational Details

Today, there are 52 One-Call centers in the United States. While these are generally state-level call centers, larger states, like California have two, while smaller states may share one regional call center, like Maine, Massachusetts, New Hampshire, Vermont, and Rhode Island. Chicago also has its own center, and while Washington, D.C. is a special district with its own rules and regulations, it is part of the Miss Utility center, which also serves Maryland. No matter where someone is across the country, the same 811 call will be routed through local phone services to the call center for their location.

As the history and evolution detailed previously explained, One-Call Centers were shaped by both local collaboration and state and federal instruction. As individual entities today, these centers are most often nonprofits and are governed by its members – utility companies – via a Board of Directors. The centers are run by professional staff, led by an Executive Director,

President, or CEO. Some One-Call Centers hire contractors who specialize in running One-Call Centers and some are self-run. When contractors are hired, all of the employees work for the contractor, except the Executive Director/President and a small staff. New Jersey is an exception to this rule, where the Board of Directors works directly with the call center contractor without a staff-level president or executive director.[75] About one-third of the states use software contractors to manage the center.[76]

Regardless of the staff or contractor mix conducting daily operations, a common denominator nationally is that the One-Call system leadership chosen by the board are professionals who are extremely passionate about damage prevention. Executive Director/President/CEO turnover is very low because of this passion, which allows them to focus on ways to improve efficiency, education, and innovation.

A striking example of how adaptable and cooperative these arrangements can be comes from across the border. There's a little-known story about Pennsylvania 811 operating Alberta One-Call's Contact Centre out of Pennsylvania for at least three years before Alberta was able to take over operations. That story alone illustrates the quiet collaboration Canada and the United States have shared on our collective and parallel damage prevention journey and it really deserves to be told. Mike Sullivan shares his thoughts here:

> Canada and the United States share a long, collaborative, and celebrated journey in safety and damage prevention – one that transcends time zones, cultures, and languages in pursuit of shared goals. As someone deeply committed to the continual improvement of the damage prevention process, I'm proud to carry forward the perspective and vision initiated by my predecessors, Scott Henley and Bob Chisholm. I'm equally humbled by the knowledge, expertise, and dedication of my colleagues and stakeholders, whom I have the privilege to represent. Indeed, 'The truth is out there,' and we'll find it far more effectively together than apart.

The business model of the One-Call centers relies on at least two primary funding streams, which we might simplify as memberships and notifications. Each utility company joins the One-Call center as a member, whether by state law mandate or voluntarily, and are required to pay a membership fee or dues – these may be several hundred or several thousand dollars per year depending on the size of the utility and the state or region. Uniquely, Utility Safety Partners in Alberta, Canada is the only call center in the world to use a Risk & Consequence formula to influence its membership fee structure, where those utilities employing best practices and technology can have reduced costs, while those failing to invest in damage-reducing strategies pay a higher rate.

The AMF is influenced by the geographic location of the buried asset and the product it carries or service it provides, the number of notifications which can be reduced by registering buried assets by polygon rather than grid and ongoing improvements in a member's damage prevention management through adoption of best or better practices.

The second funding source is through notifications. Essentially, when an excavator calls or goes online to request a locate, the One-Call center sends a notification to each utility in the designated area – often up to six companies. There is a cost to each of those notifications – often a nominal rate of around a dollar. While the excavator pays nothing, the utility company essentially pays every time someone digs around their lines. It is a small amount, and as technology and best practices help refine ticket screening, they will likely save more in the future. But as many utilities would likely agree, it is better to pay a dollar and to be forward leaning to send locators and prevent damage than risk untold millions in repairs, downtime, and associated costs. This balance between free excavator service and relatively low utility cost is important, but education remains key.

> I really want to keep this idea that the one call, 811 process is free, to not drive people away, not scare them off, because we still see a lot of damages, a high percentage of damages, where people don't even start in the process. They don't even put in that one call ticket.
>
> (Darcy Hurlock, TELUS, *ESA Town Hall*)

With each 811 Center being made up of stakeholders with buried assets, locators, and in some cases contractors, these are all the companies and people who are on the front lines with vested interest in excavation safety and damage prevention. Many of these organizations provide extensive free training, funded by 811 Center membership fees and sponsorships.

However, One-Call centers responding to the needs of the industry itself are only part of the services it provides. At the end of the day, the number one priority of the One-Call is to organize and communicate between different stakeholders to facilitate locate requests that prevent damage. That means the process must be just as responsive to professionals as it is to regular homeowners who are unfamiliar with this system.

> A homeowner, they're going to dig once every 10 years, maybe once in a lifetime. This isn't what they do for a living. So they need that extra time, that extra help.
>
> (Mike Sullivan, President, Utility Safety Partners, *ESA Town Hall*)

Driving Continuous Improvement

The 811 Centers are the heart and soul of the damage prevention industry. Because they interface with virtually every stakeholder group, they are the *common ground* in an operational day-by-day sense. The investments they make and the programs they run have a direct impact on damage prevention by expanding their reach to more homeowners, first-time excavators, and professional contractors to ensure they are using all the tools available to them. To do that effectively, they must align marketing, education, awareness, and more with basic logistics.

As 811 took hold, many of the U.S. One-Call Centers incorporated 811 into their names to capitalize on the branding and recognition and reinforce awareness campaigns. Kentucky and Indiana, followed quickly by Tennessee, were the first centers to add 811 to their names, which started the snowball rolling.[77] Today, more than 30 centers have "811" in their official name, and all of them use 811 symbols and materials for outreach.

Our practice in this book has been to use 811 Centers as a catch all within the United States and Facility Notification Centers when including international examples. We have refrained from using "One-Call" except for historical accuracy. Some industry veterans may desire a specific term they think is best. We are not seeking to reinforce old language nor preferences for new language, but acknowledge that the industry is in flux once more. While major shifts occurred in prior decades to formalize its purpose (e.g., utility protection, facility notification), its function (e.g., One-Call), and awareness efforts (e.g., 811), new dynamics are already at play. As the industry navigates the shift from primarily phone call locate requests to digital (both online and mobile application-based) tickets through the call center websites, it is once again reconsidering naming conventions to reinforce education and awareness.

The initial advantage of creating a nationwide phone number for protecting utilities has in some ways become a challenge. For two decades, the entire industry has been associated with a phone number, yet in recent years online locate requests and mobile applications have taken off, far exceeding the number of phone requests. In 2024, for instance, overall ticket requests increased from the previous year by 2.55 percent, and the breakdown of clicks to calls was more than 3 to 1. A total of 33,447,167 online locate requests were made nationwide through 811 Centers (a 6.19 percent increase from the prior year), while 10,079,850 calls were made in the United States (a 9.69 percent decrease from the prior year).[78]

The 811 logo created by the CGA has been deeply ingrained in the industry and was initially intended to get excavators to "call 811." As the world has moved to seemingly everything being online, it has been my (Scott's) experience that 811 has become a verb for many people. For many, 811 means the action of requesting a locate with no deference to whether it is a phone call or an online request. We still do this in other areas – like saying *dial* when we mean tap flat virtual buttons instead of turning a rotary on a landline phone; or saying we *called in* a food order when we placed it through the internet. If you can believe it, we used to call the pizza shop not long ago to place orders, and we now use online orders that allow us to track the pizza through the oven.

However, the use of 811 to simply mean "request a locate" has caused some confusion about which stakeholder does what in the process. These questions will be dealt with, and hopefully eliminated over time.

The CGA has adapted the 8-1-1 logo tag line over time to reflect the changing way locate requests are submitted. The CGA provides several options so the user can make their own choice.[79] As of August 2025 the logo options are (English and Spanish versions);

1 Original 811 logo with no tag line
2 Logo with "Know what's below, Call before you dig."
3 Logo with "SAFETY IS IN YOUR HANDS. EVERY DIG. EVERY TIME."

The 811 Logo is a registered trademark of the CGA. It is also seemingly ubiquitous. We almost guarantee you've seen it as a truck bumper sticker, on a billboard, or even shirts, hats, and other swag. Close your eyes and try to picture it in your head. Think you have it?

Scan the code to see how you did! And now, commit it to memory!

In the DIRT Report for 2019, CGA was exploring ways to be proactive in getting more people to make locate requests before digging. One starting point is the acclaimed CGA Best Practice Guide, which is a consensus report agreed to unanimously by all 16 stakeholder groups.[80] In the

Report, CGA included a note following Best Practice Statement 5.1 "One Call Facility Locate Request" – or the best practice of making the basic click or call to 811 – which reads:

> Practice statement: The Excavator requests the location of underground facilities at each site by notifying the facility owner/operator through the one call center. Unless otherwise specified in state/provincial law, the excavator calls the one call center at least two working days and no more than ten working days prior to beginning excavation.

The note stated: "Update Opportunity: Consider updating to reflect three-digit dialing (811) which was introduced in 2007, and that electronic notifications have become the predominant method of one call center notices."

That proposal might read something like this (own words here):

> Practice statement: The Excavator requests the location of underground facilities at each site by notifying the member facility owner/operator through the one call center. Unless otherwise specified in state/provincial law, the excavator calls ***811 or uses*** the one call center ***website to request an online locate request*** at least two working days and no more than ten working days prior to beginning excavation.

This more natural statement would have both emphasized the phone number itself, already long established, while also pointing to the increasingly common practice of web-based ticket entry. The shift to electronic requests has been underway for some time. Overall locate requests have steadily increased at least since 2010, with fax requests declining sharply and all but disappearing in recent years. Electronic requests appear to be driving the growth in total locates, especially with phone requests declining, becoming the dominant contact method between 2015 and 2018.

Per Susan Bohl, Executive Director of Oklahoma One Call System (dba OKIE811),

> Over the past seven or more years, there's been a concerted effort to move excavators to entering their own online locate requests. In return for entering their own requests and not having to have the 811 Center enter, review, and submit requests for the excavator, the excavator gets their ticket number immediately and locate notices go out much faster.

The opportunity 811 provided was unique to the United States and limited to this country's telecommunications practice and federal policy. It helped make damage prevention successful. Industry efforts to capitalize on it then created unforeseeable hurdles as technology changed. Outside of the United States, however, while there was no initial 811 bump, there was also no internet-driven phone obsoletion slump.

Clicks Not Calls

The impact of not using 811 is illustrated well with Canada's experience. There was a concerted effort to take advantage of the 811 program in Canada, but this number was already in use for non-emergency health services and was not available. This effectively disenfranchised the nation from joining the effort and uniting the majority of the continent, leaving Canada to develop and deploy its own messaging. This would ultimately birth the "click before you dig" campaign and the trademarked ClickBeforeYouDig brand, owned by Alberta One-Call Corporation.

From the early 2000s onward, Canadian damage prevention still utilized the industry standard "Call Before You Dig," and when the U.S. designated 811, nothing changed in the north. What was changing for the whole continent was increased internet and telecommunications infrastructure and higher broadband penetration. By the 2010s, the web-based economy was firmly established globally and improvements in damage prevention were increasingly happening online. In 2012, ClickBeforeYouDig.com was registered and became the Canadian version of 811, which focused the Canadian push toward online locate requests, not via the phone.[81] The website also links to U.S. 811 Centers.

Alberta was specifically committed to converting calls into online locate requests, and took the campaign a step further. In 2013, Mike Sullivan, President of Utility Safety Partners (formerly Alberta One-Call) took the bold step of removing their phone number from all marketing materials and pointing everyone to online requests. He shared that within six weeks of only promoting ClickBeforeYouDig, online locate requests jumped from 17 percent to over 60 percent.

In a follow-up study by USP in 2019, data indicated the damage rate is lower with online requests vs requests placed via the phone by 50 percent.[82] They believe the damage reduction was the result of a few key factors: First and foremost, the person submitting the online locate request typically knows precisely where they are digging, and the online process allows them to indicate the location with interactive mapping or geolocation tools, identifying the dig site with precision. Phone requests can be inaccurate, misheard, misinterpreted, or entered incorrectly by a Facility Notification Center Agent – not due to poor quality work, but the inherent difficulty in relaying information through a call. Easy to use drop down options within the online locate request environment, and access to Chat with knowledgeable Facility Notification Center Agents, significantly reduces those risks. Georgia has similarly validated that excavator-generated locate requests had lower incidents of damage than agent-entered information based on phone call requests.

Once the data was confirmed, Alberta One-Call Corporation mandated online locate requests for all Members and Contractors by "soft launch" September 16, 2019 and hard-launched January 1, 2020.[83] The four-month gap between the soft and hard launch allowed USP to assist and train those members and contractors that had not made the shift from calls to clicks. Sullivan predicts that USP will mandate all locate requests to the web in 2027, as Ontario did in March 2025.[84] It is worth noting that Ontario One Call and Info-Excavation in Quebec equally agreed with USP's analysis that Clicks reduce damages compared to Calls. In fact, when Ontario One Call conducted the same analysis, its results were even more compelling than Alberta's.

Mike Sullivan explains in simple terms his view on this analysis. "If a damage has been identified, determine whether or not there was a locate request associated with it. If yes, then determine how the locate request was submitted – by web or phone/Call or Click."

Mike's additional observation:

> Some may argue there's more to it than this (software, awareness, etc.) and they're right – there is! – and this simple analysis will point you in the direction to improve whatever it is in your system that is failing. But for an industry that relies on data to make improvements, denying, ignoring or posing arguments that disregard this data is baffling.

It is too early to pass definitive judgment on the impact of online vs called in locate request outside of Alberta because of differences in excavator habits, the specific Facility Notification Center's software, and operating procedures. Scott Crawford, President and CEO at Virginia 811, raises questions intended to improve the accuracy of communicating on this topic within the industry and to drive nuanced understanding of specific root causes:

> The question remains – were the damages tied to called in tickets due to erroneous information on the ticket or was the root cause tied to mismarks (beyond issues tied to erroneous information on the ticket), excavator practices (e.g., using mechanized equipment within the tolerance zone or blind boring), etc? How many of these damages tied to called in tickets were tied directly to the ticket as being the root cause of the damage?
>
> This conclusion, that online tickets are stronger than called in tickets, is only acceptable if the damages are directly tied to errors on the ticket. We have definitive data in Virginia suggesting that online tickets are 200 times **more likely** to contain erroneous or confusing information than a called in ticket. That is why we developed an AI auditing tool to audit online tickets at 100 percent. The assumption that the online locate request "typically knows precisely where they are digging" ignores the large number of back office individuals submitting these locate requests who have never been to the excavation area and are working off of, and interpreting, notes.

In Virginia, the jury is still out on which type of ticket leads to the fewest damages. USP's program may be an anomaly, but time will tell. Fast forward five years, and it is hard to believe that online ticket entry will not be approaching 100 percent. We believe the questions raised and technical concerns found in Virginia will work themselves out as technology and AI move forward at the speed of light.

In Australia, they have a very different system, which we will cover later in this chapter, but their original name was Dial Before You Dig (DBYD), and their national number was 1100.[85] Unlike the United States, their national phone number was implemented in the early 1990s almost immediately after the launch of their national one-call system. In the Australian system, excavators are directly requesting maps or plans of the buried utilities, not locates.[86] These maps or plans were originally faxed or mailed, but they evolved to digital delivery in the early 2000s.

> From the start, even when DBYD was government-led, utilities sent maps, not people.
>
> (Duane Rodgers, PelicanCorp)

The eventual move to all online requests, made the DBYD name disconnected from how people were using the system. With rotary – "dial" – phones becoming relics even before the 1990s, many in the younger generations have never "dialed" a phone, causing the DBYD name to become dated. As a result, DBYD was frequently used as a verb, like 811 is in the United States. In 2022, DBYD renamed their system to *Before You Dig Australia* to better reflect their mission.[87] This came with updated logos and dropping the phone number from their website address.

The name change for Australia also coincided with a major overhaul in collaboration. It occurred in conjunction with integration of the previous separate state-based DBYD organizations, which had been operating in partnership with a separate Association of Australian DBYD organization. By 2022, all the state-based organizations and the national body became a single

The Australian system modified their name and logo to match the times and create a more effective message.

national Before You Dig Australia organization, with state-based Territory Managers to represent local interests.

On top of the 52 U.S. call centers, Canada has its own 10 (with Info-Excavation providing services to Atlantic Canada and Utility Safety Partners providing services to Saskatchewan and Manitoba). While often viewed as a unit, the United States and Canada diverge significantly in at least two respects: the phone number and underlying laws. Nevertheless, our northern neighbors are often included in annual CGA DIRT report information, featured alongside panels and townhalls, and are prominent participants in conferences, research, training, and other damage prevention efforts.

In Canada, Ontario is the only province with a Facility Notification Center law, so participation in their Facility Notification Center system is almost exclusively voluntary, unlike the United States.[88] As a result there is more variation in the way the provincial systems work. Like in the United States, pipeline companies are typically required to belong to the provincial Facility Notification Center system, but other than in Ontario, no other infrastructure owners are required to be members. Most provinces do require pipeline operators to be members, so pipelines that fall outside of the Canadian Energy Regulator (CER), Alberta Energy Regulator's and BC Energy Regulator's jurisdiction are covered by these mandates.[89]

Global One-Call Snapshot

For a global perspective on Facility Notification Center systems, we interviewed Duane Rodgers, CEO of PelicanCorp, about Facility Notification Center systems around the world. Rodgers' experience is rare: he has designed, built, and run one-call systems in multiple countries, exposing him to many different systems. There is nobody in the world who has more firsthand global experience with Facility Notification Center systems. Rodgers was hired to write the original software for Melbourne One Call Service, Australia's first modern one-call service. He went on to found PelicanCorp, a company that built and supported Facility Notification Center systems across Australia, New Zealand, Singapore, Canada, the United Kingdom, and eventually parts of the United States.

The first One-Call in Australia was formed in 1984 in Perth.[90,91] This was created by a small group of utilities and was 100 percent phone-based. It also did not yet utilize modern notification software and relied upon directories, phones, and excel spreadsheets. What is now the

Australian One-Call system started in 1989 as a local government-led initiative in Melbourne, Victoria called Melbourne One Call. In the early 1990s, it expanded across Victoria and into Tasmania, prompting the name change to DBYD. By the late 1990s, the system covered all seven states and territories.[92] In 2022, the states all rolled into one organization and updated their name to Before You Dig Australia.

This timeline shows the evolution of the Australian system.

- 1984 – Western Australia: Perth One Call System established (the origin of DBYD).
- 1988–1992 – NSW, VIC, QLD, SA, TAS: State-based DBYD organizations formed with major utilities.
- 1995–1996 – ACT & NT: Adopted DBYD systems aligned with national standards.
- 2022–Present: Unified BYDA organization, no longer independent state organizations.

The Australian system operates under some very different parameters than the United States or Canadian systems. The two main differences are that in Australia:

1 The excavation notification or enquiry does not create tickets, the BYDA system sends a referral to the asset owner, who then sends plans to the excavator via BYDA. In some limited cases, asset owners will also arrange their own locators such as high pressure gas mains.[93]
2 The contractor is responsible for the locate, not the facility owner. The contractor can do the locate themselves or hire a contract locator. In some cases, the locator must be approved by the facility owner.[94]

Duane's observation about Australia: "We don't seem to have catastrophic [damages] … that might be a function of safer work or just a different climate." While not a scientific study, it does provide interesting food for thought. Chris Ross, adds his observation as someone who works for the largest telecom provider in Australia:

> The Australian model forces an excavator to understand a little more about the nature of the utilities near their site and they have more skin in the game and more ownership of the location process, which can help increase their awareness. It can also be confusing due to the complexity of plans shared, which is an issue the Australian industry is exploring currently.

New Zealand's Facility Notification Center, BeforeUdig New Zealand, was launched in 2007 by PelicanCorp.[95] The system was inspired by the Australian model, with the major exception that it is entirely private. There were no legislative mandates that contributed to the formation of the New Zealand system.[96] Excavators who place enquiries receive maps/plans from the members and then often pay for locates and manage the process directly, improving accountability and efficiency.

Rodgers says: "We started in New Zealand with no legislation – just a business case. And it worked."

PelicanCorp also founded and operates beforeUdig in Singapore which is a free online service allowing excavators, designers, contractors, and homeowners to obtain utility plans for proposed excavation sites across the city-state.[97] Users submit details (project date, scope, location, etc.), draw their dig area using map tools, and receive a list of relevant asset owners for plan collection. Utility owners then provide maps, not field markings.

Europe lacks a unified Facility Notification Center model and remains fragmented, with widely varying approaches between countries.[98] Northern Europe has more developed systems with countries like the United Kingdom, the Netherlands, Belgium, and France having functioning Facility Notification Center services or mapping request portals. In most of these countries, unlike the U.S. model, systems typically send utility plans or digital maps rather than dispatching locators to apply spray paint or flags in the field.[99]

The Netherlands operates a centralized digital Facility Notification Center system known as KLIC (short for Kabels en Leidingen Informatie Centrum), which is regulated under the Dutch Wibon Law and operated by Kadaster, the national land registry agency.[100] KLIC was the first Facility Notification Center outside of the United States, and was established in 1967 after a series of digging incidents in the northern portion of the country.[101] Users (excavators, planners, etc.) submit a KLIC request through the online portal, indicating the precise dig area using GIS or mapping tools. Utility owners automatically send digital maps of their underground infrastructure via the KLIC system. The system does not provide physical field markings. There is a per request fee which in 2025 was €20–25 per ticket for a standard request with a higher fee for urgent requests. Plans are accessed through a downloadable viewer app, and while the system is technologically advanced, it has not reduced damage rates as hoped – excavations still strike buried assets in roughly six out of every 100 requests.[102]

The United Kingdom has a hybrid system, where some utilities provide locates and others push the cost to contractors. LineSearchBeforeUdig (LSBUD) was founded by a small group of pipeline companies in 2003.[103] They provide a free-of-charge public portal for excavators to submit online enquiries and receive an instant list of member asset owners, followed by individual members supplying utility plans via automated or manual process. LSBUD now processes more than 4 million requests per year, and adoption has grown within the construction and broadband sectors.[104] The shift from individual utility systems to LSBUD also revealed hidden demand: once centralized, inquiries surged, proving that many excavators were not contacting utilities directly beforehand.[105]

In the Southern region of Europe, there are no comprehensive Facility Notification Center systems; each utility has its own process, which makes it very inefficient for contractors. This lack of any organized system would seem to be a public safety hazard, but there are no damage tracking systems to verify or dispute this assumption. Even large economies like Germany rely heavily on fragmented, utility-by-utility approaches, underscoring how uneven the landscape remains.[106]

Each country's solution is deeply shaped by culture, climate, regulation, and infrastructure maturity. Rodgers emphasizes that "there's no one-size-fits-all answer." He also feels that: "The only person who hits a line is the one holding the shovel. Help him, don't just check a box."

Technology Has, and Will Continue Having, a Tremendous Impact on 811 Centers

In 2023, Aii conducted a survey of Facility Notification Center leadership (presidents, CEOs, and executive directors) at every center in the United States and Canada.[107] While the survey was entirely anonymous, it had a 60 percent response rate. The goal of the survey and report was to help Facility Notification Center leaders by sharing their peer's practices among them. Accordingly, a full survey result was only made available to the call center leaders directly, with a public version that did not include candid comments and remarks.

Among the notable findings include:

> Call centers are implementing and using technology. From regular social media usage to data collection and analysis, 80 percent or more of respondents have embraced common or emerging technology at their centers.
>
> 71 percent of call centers responding offer a mapping tool for excavators to pre-mark their dig site.[108]
>
> Approximately half of call centers responding allow a ticket to be accessible by the excavator or attached to a locator positive response – one component of an enhanced positive response. Responding centers reported that they allow the following to be attached to a positive response or are accessible to the excavator through a portal include: electronic white-ling map (40 percent), digital photographs (37 percent), facility maps (34 percent), and virtual manifests (19 percent).[109]

When asked about ways the Facility Notification Center leaders believe they can continue to support reducing damages across the damage prevention ecosystem, their role evolving in the future, and opportunities to pursue, there was a clear common thread prominently featured across dozens of free responses: technology.

Just as these centers have been shaped by the changing technological landscape, they are resolved to take destiny into their own hands and leverage technology to shape the industry. From leaning on distributed fiber optic sensing, in the Texas 811 Guardian program, to improvements in enhanced positive response in Colorado, to use of AI to help improve ticket reviews as in Virginia, 811 Centers will reduce damage. Not only are they undertaking proactive efforts with technology, but are using technology to strategically improve awareness and outreach.

Damage Prevention Education

Public awareness campaigns encouraging excavators to request a locate will be discussed more deeply in Chapter 8, but the 811 Centers play a critical role in this process. Pipeline companies are legally required to provide safety education, and 811 centers shape their training based on the guidance of their members and boards.[110] Because of an understanding of local laws, culture, and climate conditions, 811 Centers are ideally positioned to provide education to stakeholders.

Many 811 Centers have impressive education programs including field liaisons, who provide both scheduled and on-demand education across the entire service area they serve. These liaisons are extremely passionate damage prevention advocates. In many states, the 811 Center has statewide or regional events providing free education for contractors. In our experience, if a contractor wants to supplement what they learn about damage prevention and excavation safety in their internal training programs, there are plenty of in person or online opportunities provided by 811 Centers and pipeline operators.

The CGA provides some great free damage prevention education tools to use to make it easier for 811 Centers and all stakeholders, as well as to help provide consistent messaging. The 811 Centers are the true boots on the ground in the education process and many have been doing it for approaching 50 years. The creativity, passion, and dedication shown at the local levels by 811 Centers and pipelines is truly amazing.

Globally education varies, but the level of education provided seems to be linked to the Facility Notification Center system format. The systems which are national like Australia, Canada, New Zealand, United States, and the United Kingdom seem to provide the most damage prevention education.

Conclusion

The American One-Call system is just over 60 years old. In that time, immense technological and political shifts have taken place that have caused fundamental transformation in the way the system works. From organized cooperatives to standardized phone numbers, from federal mandates to state-level variation, and eventually to the use of web-based locate requests and other technology, change has been a constant.

With each of these evolutions, the goal remained the same: prevent damage, protect workers. Untold changes will be ahead, but the unified purpose will remain. The goal of zero incidents can be realized, but it will take common ground, collaboration, communication, and increased professionalization.

Chapter 6 Resources

ESA/ACTS Now Town Halls

Chaos to Common Ground in 40 Years – A 2014 presentation at Infrastructure Resources CGA Excavation Safety Conference & Expo.

Planet Underground TV – A YouTube channel packed with damage prevention and locating videos including Locator Training, Digging Dangers, Roundtables, Industry Shorts, and more.

CGA Nine Elements of an Effective State Damage Prevention Program

PHSMA State Damage Prevention Enforcement Program Adequacy

Notes

1 De Feo, G., Antoniou, G., Fardin, H., El-Gohary, F., Zheng, X., Reklaityte, I., Butler, D., Yannopoulos, S., & Angelakis, A. (2014). The historical development of Sewers Worldwide. Sustainability, 6(6), 3936–3974. https://doi.org/10.3390/su6063936

2 Wells, B., & Wells, K. L. (2016, January 30). Illuminating gaslight. https://aoghs.org/technology/manufactured-gas/

3 Burgess, N. (2015, September 1). Benchmarks: October 9, 1865: First Successful U.S. Oil Pipeline. https://www.earthmagazine.org/article/benchmarks-october-9-1865-first-successful-us-oil-pipeline

4 GeoCorr, LLC. (2021, February 24). Oil pipeline history & how they work to distribute oil across the United States. https://blog.geocorr.com/oil-pipeline-history-how-they-work-to-distribute-oil-across-the-us

5 Pickford, J. H. (1962, October). *Underground wiring in new residential areas* (Information Report No. 163). American Society of Planning Officials. https://planning-org-uploaded-media.s3.amazonaws.com/document/PAS-Report-163.pdf

6 Federal Highway Administration (FHWA). (2017, June 27). How the Highway Beautification Act Became a Law. https://www.fhwa.dot.gov/infrastructure/beauty.cfm

7 Highway Research Board. (1971). Design and Construction of Highways in Urban Areas. https://onlinepubs.trb.org/Onlinepubs/hrr/1971/372/372.pdf

8 Pennsylvania 811. (2018, April 4). Elements of a Pennsylvania damage prevention program. https://www.pa1call.org/getmedia/4e0ef3c8-97f7-4f5a-9f4e-d5f3dda41b7d/Elements-of-a-Damage-Prevention-Program-4-4-18.pdf

9 Kuykendall, R. (1974). Need for and Application of Utility-Transportation Coordination. https://onlinepubs.trb.org/Onlinepubs/trr/1976/571/571-003.pdf?

10 Substructure Committee of Los Angeles. (1949). Manual on the Surface Traffic Interference Problem as Caused by Substructure Construction and Operation. https://babel.hathitrust.org/cgi/pt?id=uc1.b5026604&seq=5

11 American Water Works Association (1949). Construction and Traffic Safety Measures: A Compilation. American Water Works Association, 41(7), 661–674. http://www.jstor.org/stable/41235219

12 Federal Highway Administration (FHWA). (1993, June). Highway/Utility Guide. https://www.fhwa.dot.gov/utilities/010604.pdf
13 Greenwood, J. T. (1989). *One-call: The first 25 years*. Public Works Historical Society.
14 Congressional Record. (1973, July 28). Congressional Record. U.S. Government Publishing Office. https://www.congress.gov/93/crecb/1973/07/28/GPO-CRECB-1973-pt21-1-1.pdf
15 Congressional Record. (1973, July 28). Congressional Record. U.S. Government Publishing Office. https://www.congress.gov/93/crecb/1973/07/28/GPO-CRECB-1973-pt21-1-1.pdf
16 Zebe, P. K. (1987, July). An examination of outside forces damage to natural gas pipelines and damage prevention. U.S. Department of Transportation, Research and Special Programs Administration, Transportation Systems Center. rosap.ntl.bts.gov/view/dot/11384/dot_11384_DS1.pdf
17 Congressional Record. (1973, July 28). Congressional Record. U.S. Government Publishing Office. https://www.congress.gov/93/crecb/1973/07/28/GPO-CRECB-1973-pt21-1-1.pdf
18 Pearson, D. (1965, June 18). Report on Pipeline Explosion. Madera Tribune. https://cdnc.ucr.edu/?a=d&d=MT19650618.2.114&e=——-en–20–1–txt-txIN——
19 Congressional Record. (1973, July 28). Congressional Record. U.S. Government Publishing Office. https://www.congress.gov/93/crecb/1973/07/28/GPO-CRECB-1973-pt21-1-1.pdf
20 National Transportation Safety Board (NTSB). (2024, August 20). History of The National Transportation Safety Board. https://www.ntsb.gov/about/history/Pages/default.aspx
21 The American Presidency Project. (1968, August 13). Statement by the president upon signing the Natural Gas Pipeline Safety Act of 1968. https://www.presidency.ucsb.edu/documents/statement-the-president-upon-signing-the-natural-gas-pipeline-safety-act-1968
22 Pipeline and Hazardous Materials Safety Administration (PHMSA). (1968). Natural Gas Pipeline Safety Act of 1968. https://www.phmsa.dot.gov/sites/phmsa.dot.gov/files/docs/Natural%20Gas%20Pipeline%20Safety%20Act%20of%201968.pdf
23 National Transportation Safety Board (NTSB). (1973, June 7). Prevention of damage to pipelines; special study. https://hdl.handle.net/2027/mdp.39015023961702
24 Pipeline and Hazardous Materials Safety Administration (PHMSA). (1975, November). Study on current practices, technologies, problems and recommendations relating to the overall safety of Gas Pipeline Distribution Systems. https://primis.phmsa.dot.gov/rd/projects/20/
25 National Transportation Safet Board (NTSB). (1973, June 7). Prevention of damage to pipelines; special study. https://hdl.handle.net/2027/mdp.39015023961702
26 National Transportation Safet Board (NTSB). (1973, June 7). Prevention of damage to pipelines; special study. https://hdl.handle.net/2027/mdp.39015023961702
27 National Transportation Safet Board (NTSB). (1973, June 7). Prevention of damage to pipelines; special study. https://hdl.handle.net/2027/mdp.39015023961702
28 National Transportation Safet Board (NTSB). (1973, June 7). Prevention of damage to pipelines; special study. https://hdl.handle.net/2027/mdp.39015023961702
29 Federal Highway Administration (FHWA). (1974, July). Accommodation of Utility Plant Within the Rights of Way of Urban Streets and Highways. https://ia803205.us.archive.org/34/items/accommodationofu00amer/accommodationofu00amer.pdf
30 Kuykendall, R. (1974). Need for and Application of Utility-Transportation Coordination. https://onlinepubs.trb.org/Onlinepubs/trr/1976/571/571-003.pdf?
31 Kiger, W. G. (2017, April 5). Testimony before the House Consumer Affairs Committee. Pennsylvania General Assembly. https://www.legis.state.pa.us/WU01/LI/TR/Transcripts/2017_0042_0001_TSTMNY.pdf
32 Pipeline and Hazardous Materials Safety Administration (PHMSA). (1977). Effectiveness of programs for prevention of damage to pipelines by outside forces (4 of 5). https://primis.phmsa.dot.gov/rd/FileGet/81/Prevention%20of%20Damage%20to%20Pipelines%20(4%20of%205)%20140-230.pdf
33 National Transportation Safet Board (NTSB). (1973, June 7). Prevention of damage to pipelines; special study. https://hdl.handle.net/2027/mdp.39015023961702
34 Punches, D. E. (1977). Standard Color Markings for Underground Facilities. https://onlinepubs.trb.org/Onlinepubs/trr/1977/631/631-011.pdf
35 Transportation Research Board. (1988). Pipelines and Public Safety. https://onlinepubs.trb.org/Onlinepubs/sr/sr219/219.pdf
36 Dierker, B., & Rogers, O. (2024, September). 2024 Damage Prevention Report Card. https://www.aii.org/wp-content/uploads/2024/10/2024-Damage-Prevention-Report-Card.pdf

37 Miss. Code Ann. §§ 77-13-1 to 77-13-37. (2025). Regulation of Excavations Near Underground Utility Facilities. https://www.msdamageprevention.com/law/

38 Va. Code Ann. § 56-265.15:1. (2025). Exemptions; routine maintenance. https://law.lis.virginia.gov/vacode/title56/chapter10.3/section56-265.15%3A1/

39 Underground Utility Line Protection Law, 73 P.S. § 176 et seq. (2025). https://www.legis.state.pa.us/WU01/LI/LI/US/HTM/1974/0/0287.HTM

40 Miss. Code Ann. §§ 77-13-1 to 77-13-37. (2025). Regulation of Excavations Near Underground Utility Facilities. https://www.msdamageprevention.com/law/

41 Common Ground Alliance. (2024). 2023 DIRT report: Analysis & recommendations. https://dirt.commongroundalliance.com/2023-DIRT-Report

42 Damage Prevention Program, 49 C.F.R. § 192.614. (2026). Electronic Code of Federal Regulations. https://www.ecfr.gov/current/title-49/subtitle-B/chapter-I/subchapter-D/part-192/subpart-L/section-192.614

43 Damage Prevention Program, 49 C.F.R. § 192.614. (2026). Electronic Code of Federal Regulations. https://www.ecfr.gov/current/title-49/subtitle-B/chapter-I/subchapter-D/part-192/subpart-L/section-192.614

44 Utility Notification Center of Colorado. (2017). Form 990: Return of organization exempt from income tax. Internal Revenue Service. https://apps.irs.gov/pub/epostcard/cor/841042045_201712_990O_2019020116055283.pdf

45 Federal Highway Administration (FHWA). (2022, May 31). The Transportation Equity Act for the 21st century. https://www.fhwa.dot.gov/tea21/index.htm

46 United States Department of Transportation; Research and Special Programs Administration; Office of Pipeline Safety. (1999, August). Common ground study of one-call systems and Damage Prevention Best Practices. https://primis.phmsa.dot.gov/comm/publications/CommonGroundStudy090499.pdf

47 Common Ground Alliance (CGA). (2026, February). CGA Best Practices Version 22.0; Establishment of the Common Ground Alliance. https://bestpractices.commongroundalliance.com/1-Introduction/Establishment-of-the-Common-Ground-Alliance

48 United States Department of Transportation; Research and Special Programs Administration; Office of Pipeline Safety. (1999, August). Common ground study of one-call systems and Damage Prevention Best Practices. https://primis.phmsa.dot.gov/comm/publications/CommonGroundStudy090499.pdf

49 Pipeline and Hazardous Materials Safety Administration (PHMSA). (2002) Pipeline Safety Improvement Act of 2002, Pub. L. No. 107-355, 116 Stat. 2985. https://www.npms.phmsa.dot.gov/Documents/Pipeline_Safety_Improvement_Act_2002.pdf

50 Common Ground Alliance (CGA). (2025). Data Reporting & Evaluation. https://commongroundalliance.com/Membership-Engagement/Committees/Data-Reporting-Evaluation

51 U.S. Congress. (2002, July 23). Pipeline infrastructure protection to enhance security and safety act. Congressional Record, 148(101), H5273-H5288. https://www.congress.gov/congressional-record/volume-148/issue-101/house-section/article/H5273-2

52 U.S. Congress. (2006, December 29). Pipeline Inspection, Protection, Enforcement, and Safety Act of 2006, Pub. L. No. 109-468, 120 Stat. 3486. Pipeline and Hazardous Materials Safety Administration. https://www.phmsa.dot.gov/pipeline/gas-distribution-integrity-management/pipeline-inspection-enforcement-and-protection-act-of-2006-pipes

53 Pipeline and Hazardous Materials Safety Administration (PHMSA). (2021). Nine Elements of Effective Damage Prevention Programs. https://primis.phmsa.dot.gov/comm/DamagePrevention9Elements.htm

54 Pipeline and Hazardous Materials Safety Administration (PHMSA). (2007, Spring). PHMSA Focus Special Edition: One Nation … One Number … https://data.ntsb.gov/Docket/Document/docBLOB?FileExtension=.PDF&FileName=PHMSA+unveiling+of+811+nationwide-Master.PDF&ID=40364721&utm_

55 Pipeline and Hazardous Materials Safety Administration (PHMSA). (2015). Characterization of State Damage Prevention Programs. https://primis.phmsa.dot.gov/comm/sdppcdiscussion.htm?nocache=715

56 Pipeline and Hazardous Materials Safety Administration (PHMSA). (2024, September 3). Determinations of Adequacy of One-Call Law Enforcement Programs from 2023 Audits. https://www.phmsa.dot.gov/sites/phmsa.dot.gov/files/2024-09/2024-Adequacy-Map-09-03-2024.pdf

57 Congressional Record. (1973, July 28). Congressional Record. U.S. Government Publishing Office. https://www.congress.gov/93/crecb/1973/07/28/GPO-CRECB-1973-pt21-1-1.pdf

58 The American Presidency Project. (1968, August 13). Statement by the president upon signing the Natural Gas Pipeline Safety Act of 1968. https://www.presidency.ucsb.edu/documents/statement-the-president-upon-signing-the-natural-gas-pipeline-safety-act-1968

59 United States. Bureau of Surface Transportation Safety. (1973). Prevention of damage to pipelines: special study. Washington: National Transportation Safety Board. https://catalog.hathitrust.org/Record/000940443
60 Federal Highway Administration (FHWA). (1974, July). Accommodation of Utility Plant Within the Rights of Way of Urban Streets and Highways. https://ia803205.us.archive.org/34/items/accommodationofu00amer/accommodationofu00amer.pdf
61 Kiger, W. G. (2017, April 5). Testimony before the House Consumer Affairs Committee. Pennsylvania General Assembly. https://www.legis.state.pa.us/WU01/LI/TR/Transcripts/2017_0042_0001_TSTMNY.pdf
62 American Public Works Association through One-Call Systems International. (1980, November 12). One Call Systems International (U.S. Trademark Application Serial No. 73285558). https://uspto.report/TM/73285558
63 Punches, D. E. (1977). Standard Color Markings for Underground Facilities. https://onlinepubs.trb.org/Onlinepubs/trr/1977/631/631-011.pdf
64 Damage Prevention Program, 49 C.F.R. § 192.614. (2026). Electronic Code of Federal Regulations. https://www.ecfr.gov/current/title-49/subtitle-B/chapter-I/subchapter-D/part-192/subpart-L/section-192.614
65 Common Ground Alliance (CGA). (2024). One Call Systems International. https://ocsi.commongroundalliance.com/
66 Utility Notification Center of Colorado. (2017). Form 990: Return of organization exempt from income tax. Internal Revenue Service. https://apps.irs.gov/pub/epostcard/cor/841042045_201712_990O_2019020116055283.pdf
67 United States Department of Transportation; Research and Special Programs Administration; Office of Pipeline Safety. (1999, August). Common ground study of one-call systems and Damage Prevention Best Practices. https://primis.phmsa.dot.gov/comm/publications/CommonGroundStudy090499.pdf
68 Common Ground Alliance (CGA). (2026, February). CGA Best Practices Version 22.0; Establishment of the Common Ground Alliance. https://bestpractices.commongroundalliance.com/1-Introduction/Establishment-of-the-Common-Ground-Alliance
69 U.S. Congress. (2001, December 20). H.R. 3609, Pipeline Safety Improvement Act of 2002, 107th Congress (2001–2002). https://www.congress.gov/bill/107th-congress/house-bill/3609
70 Common Ground Alliance (CGA). (2025). Data Reporting & Evaluation. https://commongroundalliance.com/Membership-Engagement/Committees/Data-Reporting-Evaluation
71 Federal Communications Commission (FCC). (2005, March 10). FCC Designates 811 as the Nationwide Number to Protect Pipelines, Utilities from Excavation Damage. https://docs.fcc.gov/public/attachments/DOC-257293A1.pdf
72 Pipeline and Hazardous Materials Safety Administration (PHMSA). (2007, Spring). PHMSA Focus Special Edition: One Nation … One Number … https://data.ntsb.gov/Docket/Document/docBLOB?FileExtension=.PDF&FileName=PHMSA+unveiling+of+811+nationwide-Master.PDF&ID=40364721&utm_
73 Pipeline and Hazardous Materials Safety Administration (PHMSA). (2015). Characterization of State Damage Prevention Programs. https://primis.phmsa.dot.gov/comm/sdppcdiscussion.htm?nocache=715
74 Facility Notification Centers Association (FNCA). (2011, November). FNCA Bylaws. https://fncainc.org/wp-content/uploads/2021/11/Bylaws-FNCA-ver6-signed.pdf
75 New Jersey Board of Public Utilities. (2023). Underground Facility Protection Act: N.J.S.A. 48:2-73 et seq. https://www.nj1-call.org/wp-content/uploads/2020/04/N.J.S.A.-48-2-73-et-seq.pdf
76 One Call Concepts, Inc. (2021, October). One Call Concepts' "Utility Defenders" Have Arrived to Protect Underground Utilities. https://utility-defenders.com/wp-content/uploads/2021/10/ud-press-releasev3.pdf
77 Indiana 811. (2009, March 30). Governor Daniels Proclaims April as Safe Digging Month; Encourages Residents to Call Indiana 811 Before They Dig. https://www.in.gov/iurc/files/Governor_Proclaims_April_Safe_Digging_Month1.pdf
78 h Common Ground Alliance (CGA). (2025, August). 2024 DIRT Report. https://dirt.commongroundalliance.com/Portals/9/Common-Ground-Alliance-DIRT-Report-2024.pdf
79 Common Ground Alliance (CGA). (2025). 811 Logo Terms of Use. https://commongroundalliance.com/Forms/Common-Ground-Alliance-811-Logo-Terms-Use
80 Common Ground Alliance. (2020, October). 2019 DIRT report: 2019 analysis & recommendations. https://commongroundalliance.com/Portals/0/Library/2020/DIRT%20Reports/2019%20DIRT%20Report%20FINAL.pdf
81 Canadian Common Ground Alliance (CCGA). (2013, June 18). Canadian One-Call Centre Committee launches "ClickBeforeYouDig.com". https://www.canadiancga.com/page-776157/1321243
82 Ontario Regional Common Ground Alliance. (2024). *2023 DIRT report*. https://orcga.com/wp-content/uploads/2024/04/ORCGA-436219-Dirt-Report-2023-Web.pdf

83 Utility Safe Partners. (2022, August 17). Click Before You Dig for Contractors. https://utilitysafety.ca/wheres-the-line/click-before-you-dig/contractors/
84 Ontario One Call. (2024). Transitioning all Homeowner Locate Requests to Online Submissions Starting March 28, 2025. https://ontarioonecall.ca/news/transitioning-all-homeowner-locate-requests-to-online-submissions-starting-march-28-2025/
85 Before You Dig Australia (BYDA). (2025, August 26). Our history. https://www.byda.com.au/about/
86 Department of Infrastructure, Transport, Regional Development, Communications, Sport and the Arts. (2025). Before You Dig Australia. https://www.infrastructure.gov.au/department/media/publications/before-you-dig
87 Before You Dig Australia (BYDA). (2022, May 4). Innovation through transformation. https://www.byda.com.au/2022/05/04/latest-news/
88 Canadian Common Ground Alliance (CCGA). (2025). Committee Work. https://www.canadiancga.com/page-790955
89 British Columbia. (2026). *Pipeline regulation*, B.C. Reg. 281/2010, under the *Energy Resource Activities Act. BC Laws*. https://www.bclaws.gov.bc.ca/civix/document/id/complete/statreg/281_2010
90 Before You Dig Australia (BYDA). (2025, August 26). Our history. https://www.byda.com.au/about/
91 Before You Dig Australia (BYDA). (2025, August 26). Our history. https://www.byda.com.au/about/
92 Before You Dig Australia (BYDA). (2025, August 26). Our history. https://www.byda.com.au/about/
93 Before You Dig Australia (BYDA). (2025) General FAQs. https://www.byda.com.au/faqs/
94 Lee, N. (2023, April 4). WorkSafe Construction Guidance. https://www.ccfvic.com.au/worksafe-construction-guidance/
95 Pelicancorp (BeforeUDig NZ). (2020). Ultrafast Fibre leading New Zealand's fibre revolution. https://cdn.pelicancorp.com/pdfs/PCNZ-bUd-UltraFast-Fibre-Case-Study-Final.pdf
96 BeforeUdig NZ. (2019). NZ Law & Regulations. https://www.beforeudig.co.nz/nz/safety/nz-law-regulations
97 PelicanCorp. (2024). Terms of Use: beforeUdig Service. https://cdn.pelicancorp.com/pdfs/b4sg_Terms.pdf
98 Nijman, J.-W. (2023, November 20). One call in Europe. https://actsnowinc.com/global-811-magazine/one-call-in-europe
99 UK Power Networks. (2025). Request plans showing where electricity cables are. https://www.ukpowernetworks.co.uk/safety-equipment/advice/request-plans-showing-where-electricity-cables-are
100 Koppens, B. (2014). KLIC the Dutch one call system. https://www.concawe.eu/wp-content/uploads/2017/01/s3-4_summary_klic_the_dutch_one_call_system-2014-01257-01-e.pdf
101 Koppens, B. (2014). KLIC the Dutch one call system. https://www.concawe.eu/wp-content/uploads/2017/01/s3-4_summary_klic_the_dutch_one_call_system-2014-01257-01-e.pdf
102 Nijman, J.-W. (2023, November 20). One call in Europe. https://actsnowinc.com/global-811-magazine/one-call-in-europe
103 Broome, R. (2018, July 24). Increasing the visibility of safe digging plans in the UK. https://actsnowinc.com/global-811-magazine/increasing-the-visibility-of-safe-digging-plans-in-the-uk
104 LSBUD. (2025, July 15). Digging up Britain 2025. https://downloads.safedigs.co.uk/lsbud_website/pdfs/lsbud-digging-up-britain-2025-report.pdf
105 LSBUD. (2023, October 24). UKPN UK Power Networks Case Study. https://lsbud.co.uk/ukp/
106 BIL. (2025). BIL cooperative. https://bil-leitungsauskunft.de/genossenschaft/
107 Dierker, B. (2024, June). Damage Prevention One-Call Center Survey. Alliance for Innovation and Infrastructure. https://www.aii.org/wp-content/uploads/2024/06/One-Call-Survey.-Public-Results.pdf
108 Dierker, B. (2024, June). Damage Prevention One-Call Center Survey. Alliance for Innovation and Infrastructure. https://www.aii.org/wp-content/uploads/2024/06/One-Call-Survey.-Public-Results.pdf
109 Dierker, B. (2024, June). Damage Prevention One-Call Center Survey. Alliance for Innovation and Infrastructure. https://www.aii.org/wp-content/uploads/2024/06/One-Call-Survey.-Public-Results.pdf
110 Public Awareness, 49 C.F.R. § 192.616. (2026). Electronic Code of Federal Regulations. https://www.ecfr.gov/current/title-49/subtitle-B/chapter-I/subchapter-D/part-192/subpart-L/section-192.616

Appendix

State	*Current Center Name*	*Previous Names*	*Founded*
Alabama	Alabama 811	MISS ALL, Alabama Line Location Center	1975
Alaska	811 Alaska Digline		1988
Arizona	Arizona 811	Blue Stake Center, Arizona Blue Stake	1974
Arkansas	Arkansas One-Call, Arkansas 811	Arkansas One Call	1978
California	Underground Service Alert North (USA North)		1976
	DigAlert	Underground Service Alert South	1976
Colorado	Colorado 811	Utility Notification Center of Colorado, Blue Stake	1986
Connecticut	Call Before You Dig (CBYD)		1977
Delaware	Delmarva 811	Miss Utility of Delmarva	1978
D.C.	Miss Utility		1978
Florida	Sunshine 811	Florida One Call, Call Candy, Call U.N.C.L.E.	1993
Georgia	Georgia 811	Utilities Protection Center	1974
Hawaii	Hawaii 811		1988
Idaho	Digline Inc		1990
	North Idaho 811		1990
	Kootenai County One Call		1990
Illinois	JULIE (JULIE Before You Dig)	Joint Utility Locating Information for Excavators	1974
	811 Chicago	DIGGER	1974
Indiana	Indiana 811	Indiana Underground Plant Protection Service, Had-Help, Be-A-Ware, Ruff Dig-In Service, 90-90 Dig In	1979
Iowa	Iowa One Call	UNDERGROUND PLANT LOCATION SERVICES, INC,	1980
Kansas	Kansas 811	Kan-U-Dig-It	1978
Kentucky	Kentucky 811	Kentucky Underground Protection, BUD (Before-U-Dig)	1987
Louisiana	Louisiana 811	DOTTIE (Dial One Time To Inform Everyone)	1975
Maine	Dig Safe		1981
Maryland	Miss Utility	Miss Utility of Maryland	1978
	Delmarva 811		1978
Massachusetts	Dig Safe		1981
Michigan	MISS DIG 811	MISS DIG	1974
Minnesota	Gopher State One Call		1987
Mississippi	Mississippi 811	Mississippi One Call System	1984
Missouri	Missouri 811	Missouri One Call System, TO BEGIN	1985
Montana	Montana 811		1998
Nebraska	Nebraska 811	Digger's Hotline, One Call Covers All,	1994

(Continued)

State	*Current Center Name*	*Previous Names*	*Founded*
Nevada	Underground Service Alert North (USA North)	Can You Dig It	1976
New Hampshire	Dig Safe		1981
New Jersey	New Jersey One Call	Garden State Underground Plant Location Service	1975
New Mexico	New Mexico One Call, New Mexico811	New Mexico One Call, Blue Stake	1990
New York	UDig NY	Underground Facilities Protection Organization, Dig Safe New York 811	1970 (origin 1964)
	New York 811	New York City One Call	1990 (origin 1964)
North Carolina	North Carolina 811	Utility Notification Center of North Carolina, UNNC, Utilities Locating Co., Inc. "ULOCO"	1978
North Dakota	North Dakota 811		1995
Ohio	Ohio 811	Ohio Utility Protection System, OUPS, United Utilities Protection Service	1972
Oklahoma	OKIE811	Call Okie, Oklahoma One Call Center	1979
Oregon	Oregon 811	Oregon Utility Notification Center	1995
Pennsylvania	Pennsylvania One Call System Pennsylvania 811	One Call Pennsylvania One Call System	1972
Rhode Island	Dig Safe		1981
South Carolina	South Carolina 811	PUPS (Palmetto Utility Protection Service or Palmetto Utility Location Service)	1978
South Dakota	South Dakota 811	South Dakota One Call Notification Board	1993
Tennessee	Tennessee 811	Miss Locate, Tennessee One-Call System	1983
Texas	Texas 811	Texas Underground Facility Notification Corporation Texas Excavation Safety System, TESS, Texas One Call	1984
Utah	Blue Stakes of Utah 811	Blue Stakes of Utah Utility Notification Center, Joint Utilities Protection Center	1974
Vermont	Dig Safe		1981
Virginia	Virginia 811	Miss Utility of Virginia, Miss Utility of Lynchburg, Roanoke Valley Underground Location Service	1976
Washington	Washington 811	Utilities Underground Location Center	1978

(Continued)

State	*Current Center Name*	*Previous Names*	*Founded*
	NW 811	North West Utility Notification Center	1998
	IEUCC 811	The Inland Empire Utility Coordination Council	1990
West Virginia	West Virginia 811	Miss Utility of West Virginia, West Virginia One Call System, Cable Protection Bureau	1980
Wisconsin	Diggers Hotline		1976
Wyoming	Wyoming 811	One Call of Wyoming, Call-In-Dig In Safety Commission	1993

Province/ Territory	*Center Name*	*Year Founded*
Alberta	Utility Safety Partners (formerly Alberta One-Call)	**1984**
British Columbia	BC 1 Call	*1994*
Manitoba	Click Before You Dig Manitoba	*2013*
New Brunswick	Info-Excavation NB	*City of Saint John, NB provided services by Info-Excavation starting in 2002. Services expanded to the remainder of the province and Atlantic Canada (Nova Scotia, NFLD, Prince Edward Island) in 2016*
Newfoundland & Labrador	Info-Excavation NL	*2016*
Nova Scotia	Info-Excavation NS	*2016*
Northwest Territories	NWT One-Call (via Enbridge)	*No Territorial One-Call Service*
Nunavut	Nunavut One-Call (via portal)	*No Territorial One-Call Service*
Ontario	Ontario One-Call System	**1996 (legislation secured 2012 and ON1Call re-introduced in 2014 as a new entity)**
Prince Edward Island	PEI One-Call (Info-Excavation)	*2016*
Quebec	Info-Excavation QC	*1993*
Saskatchewan	Sask 1st Call	*2003*
Yukon	Yukon One-Call (via portal)	*No Territorial One-Call Service*

Historic Photo Gallery

This gallery shows images from the 1890s through the early 1900s, and up to 1962 just before the modern One-Call Center was formed. From these early roots, the industry has grown into the form seen today. But importantly, these historic images from San Francisco, CA, and Boston, MA, illustrate that the problems have remained consistent – more infrastructure is needed, with the need for safety training growing in equal proportion.

Workers at rest in gas pipe trench; passersby on sidewalk. Boston Gas Light Company.

DOI: 10.1201/9781003796619-8

Employee working with gas pipe in trench; view of furniture store and passersby. Boston Gas Light Company.

View of gas pipes in trench; buildings, horse-drawn cart, and passersby in distance. Boston Gas Light Company.

Distribution Department, supply pipe lines, Section 8, Chestnut Hill Avenue, trench for 60-inch water main in rock, showing Boston Water Works 16-inch main relocated, Brighton, MA, November 5, 1909.

Distribution Department, Low Service Pipe Lines, pumping water from cellars after a break in the 30-inch main, Boylston Street at Boylston Place; Cameron Street, Brookline, MA, February 14, 1917.

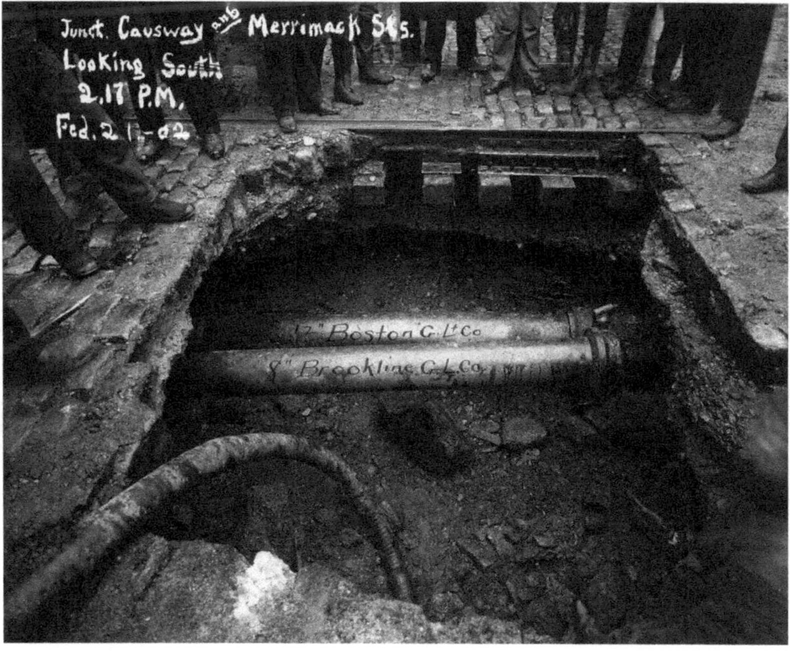

View of 12- and 8-inch gas pipes in trench. Junction of Causeway and Merrimack Streets, Looking North, 2:17 pm Boston Light Company.

Trench along Sixth Avenue and Hugo Street, looking south. From the San Francisco History Center, San Francisco Public Library.

"City Pipe System – Raised 16″ main on south side, Market Street, opposite Spear Street, view west." From the San Francisco History Center, San Francisco Public Library.

35″ Pipeline Construction Adjacent to Car Tracks, Bryant St. 16th St. to 17th Streets. From the San Francisco History Center, San Francisco Public Library.

Trench being dug along the side of Masonic Avenue, near Golden Gate Avenue. From the San Francisco History Center, San Francisco Public Library.

Distribution Department, Low Service Pipe Lines, part of Section 1, vertical curve and pipes under Boston Water Works mains, station 3+26, from the east, Brighton, MA, December 3, 1897.

"Main Break – Men at Work; Harrison street, near 17th street."

Chapter 7

Damage Prevention Fever Is Contagious

A Case Study in Passion

Kim was just looking for a job – anything that would pay the bills. She had been diligent in school, but never found that one big thing that drove her. She worked at a coffee shop throughout college and interned for a nonprofit the summer before she graduated. For the last six months, she was running out the clock on her parent's welcome at home and really hoping something would click.

Her communications degree was not worthless, but it didn't exactly scream "hire me" either. She was accepting that she would take the next job that came along, work there a few years, then shuffle around until maybe getting married and having kids. Kim was young, so a paid internship was just as good as a job. She was really casting a pretty wide net at this point.

Then she got an offer. Truth be told, she had no idea what the company did, but they needed a communications intern for a big summer campaign. With her December graduation, she knew the summer would be competitive with students and recent graduates, so she jumped at the opportunity.

Her job was video editing and some light graphic design. In short, she was going to be a social media intern. Not the calling she dreamed of, but it was something to do and she'd get $12 an hour for it.

Kim's first assignment was to edit an interview the company had recorded the week before. It was a 25-minute conversation between some old guy and some other old guy. She didn't even listen to the content, she just glazed over and listened for the "ums" and "ahs" to cut out for the final video. As she couldn't help but pick a few things up, she realized they were talking about digging – although truth be told again, she heard "excavation" a number of times and thought the first old guy may have been a paleontologist or something. A few minutes further into the recording, and all the talk of pipelines seemed to eliminate that possibility.

She handed over the cleaned up video to her boss and went about making a thumbnail for the video. Kim had been trained by years of Instagram, TikTok, and YouTube to know that people only click on something they think is going to be super interesting. Sometimes the art of a good video thumbnail was misleading the audience into thinking it would be exciting or even controversial. She didn't know the vibe her boss had, so she made a few options.

The first was an image of the first old dude standing over an open pit with a dinosaur skeleton inside with giant block letters saying "HE FOUND WHAT?" She had a hunch that even though her erroneous understanding was totally off base that maybe someone else would hear "excavation" and think of this. The other options included the two men's heads facing one another with a colorful background and text saying "Listen to the Experts" and one that was simply a black screen with small print saying "Click to reveal what's inside."

DOI: 10.1201/9781003796619-9

It was her first assignment, these were all trial balloons. She fully expected her boss would redirect and clarify the assignment with gentle and constructive feedback. That isn't what happened.

Kim sent the image files to her boss around 11 am that Wednesday morning and waited around clicking through the company website for about an hour. At lunch, she left the office for some fresh air and a cup of noodles. She didn't put another thought to the task and when she got back to her desk, started looking at old social media posts the company had made over the years. She wasn't impressed with the average of three or five likes on each post, with maybe a good one getting 12 or 15 reactions. The more she read, the more she learned. It turned out, this company was protecting pipelines from blowing up. Seemed neat to Kim. The rest of the week, she was shadowing someone at the company going across town meeting with people, so she was busy taking notes and taking pictures.

When she got into the office on Monday, her boss was standing in her office. Kim's stomach sank. She wasn't sure what her boss was doing there, but she didn't think it could be good. "Kim, I want to talk to you," her boss said in an even tone without giving anything away.

When they were both sitting, her boss asked Kim something critical: "Have you seen our social media?"

Kim wasn't sure what this meant. Did she mean at all – *yes, she spent time last week reviewing!* Did she mean a specific platform – *okay, boomer, what site do you mean?!* Did they get a bad review – *what am I supposed to do about that?!*

Her boss loved the images Kim had made for the thumbnails and actually used all three of them for different posts. They were a hit! Over the weekend, they had generated over 100,000 impressions and tons of new people were clicking on the videos and other posts. She told Kim that she had a knack for marketing. The dinosaur image alone received 70 likes on Facebook and another 32 on LinkedIn. Her boss praised her out of the box thinking.

This energized Kim, and she was beet red hearing the news. When her boss left the room, she logged online at a lightning pace to see for herself. But her mind was also racing with new ideas. She wanted to figure out exactly what the company did so she could share more information and do a better job.

The first thing she did was to go back and actually listen to the video she had already edited. When she did, she learned that one of the old guys had saved an entire community from a potential explosion. The other old guy was an emergency responder the day of the incident and he was giving details about how an elementary school and dog grooming business were nearby, which were both protected by the man's actions. She got a little lost in the details about digging, trenches, OSHA, methane leaks, and other jargon, but she understood that a pipeline had been struck, it could have killed a lot of people, but nobody was hurt.

That take away really affected Kim. She wanted *to know even more*, but she also wanted *even more people to know*! The first thing she did was make an all-new video thumbnail (now knowing just what the video was really about) that included a fireman running from a blast with a puppy and toddler in each arm. She sent this to her boss with a note, "I made another because I think everyone should see this video! Let's share it again."

- - - - - - - - - - - -

Once you realize what the mission of damage prevention is all about, it tends to get into your blood. And once it takes hold, it is hard to leave the industry. Benjamin will tell in this chapter how Scott got hooked in the early 80s, after going to work at Carsonite selling fiber-reinforced marker posts. Still in his mid-20s, it was becoming more than a job, but an interest. Then something really sealed the deal – Scott started attending the International One Call Symposiums and was immersed in damage prevention and the passion of the 811 Center operators. From that point on, it became a calling for Scott.

After almost eight years at Carsonite, with a good deal of experience and his first patent under his belt, Scott's position was eliminated – yes, he got fired! But it was too late to keep him out of the damage prevention industry, he was a true believer. So, with no one to employ him, and still a young and enterprising man in his 30s, he decided to take matters into his own hands: he started Rhino.

I hope that as you read on, some of the work Scott did in partnership with the industry will light a similar fire in you and stimulate ideas for how you too can get involved.

Rhino Damage Prevention Systems

Rhino manufactured pipeline and cable route marking products, but the real focus was helping customers prevent damage to their facilities. This passion for damage prevention led to Scott writing a recurring column, *Identification Required*, for Underground Focus Magazine for many years, as well as maintaining an ongoing connection with 811 Centers.

After the Common Ground Alliance (CGA) was formed, Scott became an active member of the CGA Educational Programs and Marketing Committee, including serving four years as co-chair. In 2007, Scott received the Jim Barron Award from the CGA. This award "is given to an individual(s) or organization(s) that may or may not be a member of the CGA but has shown true dedication to efforts that enhance underground damage prevention … and has made a positive impact on the damage prevention community."[1] Rhino had other team members participate on the CGA Committees, including Chris Thome who was also a co-chair of the CGA Education Committee and was given the Ron Olitsky Award in 2017, given to the CGA member whose dedication and service to the CGA went above and beyond the call of duty during a given year, and he/she might be considered the "most valuable player."

Rhino's dedication to supporting the CGA included providing employees the time to travel to five committee meetings a year (three days each with travel) and covering all the travel costs. Many people in the industry support the CGA in this way, but most own infrastructure or are 811 Centers. As a supporting but not direct stakeholder, Rhino's commitment reflected a genuine dedication to the shared goals of damage prevention.

The passion for damage prevention was company-wide, as was evidenced by active participation in the National Safe Digging Month (NSDM) and 811 Day campaigns and initiatives by many team members – more on those in the next chapter. Damage prevention, and eventually 811, was everywhere at Rhino. When 811 was announced in 2006, Rhino immediately started advising their customers to include 811 and the 811 logo on all their warning decal and sign artwork to be ready for the 2007 launch.

Rhino installed a large silo (like a grain silo) for their resin at their Plant in Waseca, Minnesota, a small town of 9,000 people. This silo faced the street, so Rhino had a huge 811 logo decal made and installed on the silo so everyone in the small town would see the logo and the tag line, *Know what's below, Call 811 before you dig*. No one could miss the message! Rhino stuck with this theme by installing a large 811 logo on their truck that drove back and forth between facilities in Waseca. The 811 logo was the dominant image on the truck, even over and above the Rhino logo.

Rhino's manufacturing facility in Waseca, MN.

Rhino's commitment to the cause extended to developing original initiatives like the 811 Run and 811 Mudder. Those initiatives and their dedication to the Minnesota Regional CGA were all organic engagements that were the result of the passion for damage prevention that was in the company DNA.

How did it get into the DNA? That's how Scott founded it. A safety and damage prevention culture from the top down was essential, and it permeated everything they did. But Scott had too much damage prevention culture for one man to hold, and even one company could not contain it. So other initiatives flowed out from him.

Excavation Safety Alliance (ESA)

Scott's most ambitious projects reached across the industry and, eventually, internationally. A marker company was a slice of the picture, and helped industry stakeholders improve their damage prevention efforts, while making the public safer at a local level, but to help the entire industry grow and improve, Scott embarked on two parallel missions: publishing and conferencing.

To tell this story, there are a number of entity names to sort out. While they changed monikers over the years, the passion and purpose remained: preventing damage, promoting safety, and elevating the industry to new heights.

Scott founded Underground Focus Events in 2003 to provide damage prevention education through conferences and media. As it stretched and grew over the years, and to respond to evolving needs within the industry, it became Infrastructure Resource (IR) and then Excavation Safety Alliance (ESA). The ESA mission was to "Save Lives through Education and Collaboration." We will use ESA going forward to refer to this business, even when anachronistic, because it was the company name at the end.

ESA was a separate company from Rhino, but they shared office space and the same passion. This synergy in their missions and physical proximity enabled them to team up on many initiatives. ESA was constantly innovating and pushing new ideas to promote damage prevention and excavation safety. Not all the initiatives took hold at the time, but some of them may be a perfect fit for today as times have changed. The beauty is that many of the initiatives in this and the following chapter are truly evergreen and can simply be pulled off the shelf by any stakeholder at any time!

Company Name	*Underground Focus Events, LLC*	*Infrastructure Resources, LLC*	*Excavation Safety Alliance, LLC*
Years	2003–2005	2005–2023	2023–2024

- Underground Focus Events changed to Infrastructure Resources (IR) to better reflect the broad services and products provided.
- IR transitioned to the ESA because the company was very focused on both damage prevention and excavation safety. Virtual events like ESA Town Halls were very popular outgrowth. All of the company's activities were accomplished through alliances of stakeholders who had the same priorities.

The largest and farthest reaching work ESA undertook every year was putting on a damage prevention conference. This was not just a small trade show or committee meetings, but a dynamic event with speakers, educational sessions, exposition hall, product demonstrations, and more. This reflected a maturing industry – stakeholders coming together around their shared

The company brand was modified over the years to better reflect the mission.

goal of preventing damage, regardless of their specific job title or company, be it excavation, construction, locating, utility owning, and more.

Times changed, and so did the conference name. Conferences typically come in two varieties – they are founded, owned, and operated by conference producers or they are owned by industry associations or trade groups. When they are owned by associations/trade groups, they take on the financial risks and either have staff that are dedicated to event production, or they hire an event production company to produce the event. Scott did a little of both.

A founder but not owner of the *Damage Prevention Convention* in 1997, he helped bring the industry together, but knew more was needed. Scott started this conference even before forming his own company, because he recognized the need, but soon initiated the media company and new conference. The original conference put on by what became known as ESA, Underground Focus Academy, brought together damage prevention and excavation safety sessions with a trade show along with live outdoor demonstrations. In 2006, the ESA agreed to host the CGA annual member meeting at their event providing a free venue for the CGA membership and committee meetings. As active supporters of the CGA, ESA offered to add the CGA name to the event name in order to help build the CGA brand. ESA kept the same conference, trade show, and live outdoor demo format, but in 2006 the Underground Focus Academy was renamed the "CGA Excavation Safety Conference & Expo" for a number of years.

As interest from outside the United States built, it became clear that this was no longer a domestic gathering, but a truly international one, with attendees from nine countries. In its final form, this conference was named the "Global Excavation Safety Conference." The global interest in damage prevention morphed into ESA launching the first-ever damage prevention conference in Australia.

In 2018, ESA partnered with PelicanCorp, an Australia-based damage prevention leader to produce a groundbreaking event, the Oceania Damage Prevention Conference, which drew stakeholders from Australia, New Zealand, and even a few from North America. This was the first global event that was attended by stakeholders from every group. Scott remembers vividly the excitement of the attendees who were talking about ideas and solutions with all the stakeholder groups in one place. He fondly remembers that, “Seeing the enthusiasm and excitement of all these people coming together for the first time was one of the highlights of my career.”

- Scott founded the Damage Prevention Convention in 1997 with Underground Focus Magazine and Champion Productions, which owned the event.
- In 2003, Scott partnered with Underground Focus Magazine to create Underground Focus Live!, an outdoor demo fair held at Staking University in Manteno, Illinois. After the first year of the event, Scott purchased the rights to the event from Mike Parilac, the owner of Underground Focus Magazine and Staking U.
- Originally an outdoor demonstration and expo, Underground Focus Academy was the first step to forming the seminal industry conference in the United States for over two decades. This evolved into the Excavation Safety Conference & Expo, which quickly added “CGA” to its name to help promote the member-driven alliance and advance 811 messaging.
- Outside of the United States, Scott wanted to ensure a truly global impact to the life-saving mission of damage prevention. Together with PelicanCorp in 2018, the *Oceania Damage Prevention Conference* came into being.
- CGA was dropped from the name in 2021, when the CGA launched their own new conference.
- Both ESA and its conference were shuttered in mid-2024 as a result of the CGA launching a new event in 2021 to compete with ESA’s long-established conference and show. Many of the same activities and initiatives are now carried on by ACTS Now.

Outside of conferences, the flagship product of ESA was its industry magazine. While many magazines touch on topics related to pipelines, construction, and even locating, there was a need for an industry-wide publication dedicated to all stakeholder groups that enabled cross-communication and collaboration, for leaders to speak to their peers through articles, and for individuals to constantly see educational content, case studies, new product advertisements, and more.

Originally named “Damage Prevention Professional Magazine,” this evolved as well in response to industry feedback, trends, and the need for a truly resonate message with the mission. The final version of Scott’s publication was named “Excavation Safety Magazine.”

Magazine Name	*Damage Prevention Professional Magazine*	*Dp-Pro Magazine*	*Excavation Safety Magazine*
Years	2010–2019	2019–2023	2023–2024

- The print circulation of this quarterly magazine was 15,000 along with 40,000 digital editions.
- The magazine name was shortened to the DP-Pro in 2019 because Damage Prevention Professional did not resonate with the intended audience, unfortunately.

The live education events changed and expanded to meet the needs of the industry.

- The 2023 change to Excavation Safety Magazine (ESM) was linked to the broader focus on both damage prevention and excavation safety. This also tied into the popular ESA Town Halls, ESU, and the new company name.
- The ESM mission was taken over by ACTS in 2024 and merged with their state specific magazines into the *Global 811 Magazine*, which continues to be a quarterly magazine with a print circulation of more than 85,000.

To fulfill ESA's mission to "save lives through education and collaboration," a host of multimedia products and programs were created by the ESA team. These *infrastructure resources*

Damage Prevention Magazine Evolution

The magazine's name was updated to attract more readers.

helped serve the *damage prevention professionals* in many ways. Here are a number of initiatives to show the breadth of activity Scott and his team led within the industry. Many of these are expanded upon or referenced in the following chapter, while many serve as complements to ideas and programs run by others you'll read about soon. Presented in a skimmable format below, we hope you'll take the time to read through some of the details as well.

Seedlings of a Safer Industry

Scott was a bit like a seed spreader, casting ideas far and wide. Many took root and grew to great effect. Taken together, these initiatives – alongside the work of others in the industry, explored in the next chapter – represent a thriving field with a demonstrable impact on safety.

Excavation Safety Guide (ESG) 2005–2024 – Over 10 million in print

- An annual guide packed with educational content and resources to help provide excavators all the information they need to dig safely. Custom editions of the guide are produced annually for both PA811 and PAPA, as well as a handful of other 811 Centers, which would periodically have custom editions done for them. PA811 does an annual mailing to contractors

A quick snapshot of the creative initiatives created by the Rhino & ESA teams. Andy Prins (Rhino) created the Minnesota 811 Run, 811 Mudder, Global Locate Masters, Global GPR Congress, and ESA logos.

and passes them out at their Safety Days which take place across the state. PAPA mails over 500,000 a year to contractors across the United States with complete coverage in 13 states. The ESG program was taken over by ACTS in 2024.

Excavation Safety University 2008–2024

- ESA created specific training programs for damage investigation and locating. These programs combined videos with field guides that could be used in a classroom or in the field when a question came up.

Damage Prevention Hero 2010–2025

- A highly respected quarterly and annual award program. This program honors someone who has gone above and beyond their official job duties to promote damage prevention and excavation safety. The primary goal of the program is to recognize the many unsung heroes in the industry. A full page story is done on each hero along with recognition at the annual conference where the annual winner is announced, based on a vote by everyone in the industry. Showcasing these heroes motivates others to step up and make a difference. This program was taken over by ACTS in 2024.

Damage Prevention Professional 100 (DPP100) 2011–2012

- A program intended to create damage prevention metrics that any stakeholder could pursue, a bit like the Good Housekeeping Seal of Approval. Reaching a score of 100 signified that the stakeholder had a solid damage prevention program. The standards were developed by each stakeholder group, specifically for their stakeholder group. They were also being refined to make achieving the score of 100 possible regardless of the stakeholders size. Customizing the requirements leveled the playing field so the standards were reasonable for each group. The goal was to create a voluntary system of metrics that could be a public barometer for a company's commitment to damage prevention. The concept was well received, but ESA did not have the bandwidth to get it completely off the ground.

811 Run 2013–2025

- This public awareness program was created by the ESA and Rhino teams and passed on to the CGA. The runs achieved three main objectives: (1) public awareness of 811 through all the marketing of the run, (2) educating kids with hands-on events at the runs, and (3) creating 811 advocates of all the runners who participate.

Canadian Common Ground Alliance Damage Prevention Symposium (CCGA DPS) 2013–2020

- The CCGA decided to launch a damage prevention conference in 2013. The CCGA wanted ownership, so they took on all the financial responsibilities. Since many Canadians had attended ESA conferences over the years, and they liked what they saw, the CCGA hired ESA to produce the CCGA DPS for the first seven years.

811 Mudder 2014–2017

- Another public awareness event created by ESA and Rhino to promote 811 participants. A play book was created and passed on to the CGA.

Locator Safety & Awareness Week (LSAW) 2015–2025

- An annual recognition week designed to bring attention to the incredibly important job locate technicians do. ESA created free campaign tools for all stakeholders to use to show appreciation for locators during LSAW.

Excavation Safety Alliance videos and town halls 2021–Present

- Video series, Ask the Expert, Difference Makers, a SUE education series.
- Monthly virtual Town Halls. This program was taken over by ACTS in 2024.

Pipeline Ag Safety Alliance (PASA) 2015–Present

- Educating the educator is much more powerful than trying to educate every excavator. Through participation in the Minnesota Regional CGA, Chris Thome (Rhino) met a county Ag Agent and learned that there were over 3,000 agents in the United States. These agents' full-time job is to educate farmers and ranchers on safety and productivity. Chris thought that if these agents were educated on damage prevention they would pass the training on. Even though Chris was Rhino team member he spent countless hours helping ESA create and launch PASA which is a collaborative public awareness program that teams up pipeline operators and the National Association of County Agricultural Agents (NACAA). PASA educates the County Ag Agents, who then educate farmers and ranchers in nearly 3,200 counties across the United States. This program was taken over by ACTS in 2024.

Emerging Damage Prevention Leaders (EDPL) 2019–2021

- Like all industries, educating newcomers, and helping to get them energized, is a challenge in damage prevention. The EDPL mission was to attract, educate and connect the damage prevention professional leaders of tomorrow to accelerate their impact toward the industry's common goals. The program also attempted to pair seasoned industry veterans with professionals entering the industry. The cross-mentoring would help pass on institutional knowledge and spark new ideas brought to the table by the rookies. This program was well received but it was difficult to spread the word. With all the virtual tools available today it may be time to reignite a program like this, especially as the Damage Prevention Certification program takes shape (covered in Chapter 11).

Damage Prevention Week 2019–2024

- Damage Prevention Week™ brings together major industry stakeholders to discuss new ideas. Leaders have the invaluable opportunity to discuss ways to prevent damage to both underground and overhead infrastructure and improve safety in the excavating community.

The goal is to create another week on the calendar where all stakeholders can promote damage prevention and safe excavation. The week was linked to the Global Excavation Safety Conference. This link allowed for DPW to be promoted with the Global ESC and provide a place for stakeholder groups to meet.

Global GPR Congress 2021–2024

- This annual meeting brought together GPR professionals (ground penetrating radar) from around the world to facilitate discussions around how GPR could be used for damage prevention, and related applications. Almost 300 GPR professionals from around the world attended this virtual event in 2024. This was a joint venture with Bigman Geophysical.

Global Locate Masters 2022–2024

- A unique competition pitting the skills of top locate technicians from around the world against UTTO's VR simulators. Held in front of a live audience of peers and industry leaders, Global Locate Masters evaluated their performance against other competitors and recognized, rewarded, and shined a spotlight on the best of the world's utility locating professionals. The VR format projects the locate live on a screen that the audience can watch creating respect for the profession along with a better understanding of how the locate process works. This was a joint venture of ESA and UTTO.

Notification Center Board Forums (Virtual) – 2023

- As the conference producer, Scott heard from 811 Center Board members that they wanted the opportunity to network with Board members from other Boards, so ESA created a live Board Member Forum. The live forum was such a hit, that ESA began holding a quarterly Board Member Forum between the conferences. This provided a place for Notification Center Board Members to meet and discuss issues and ideas.

Industry Summits

- Each stakeholder group has unique factors and issues that impact their damage prevention programs. Contractors have different challenges than water utilities which have different hurdles than electric utilities, etc., so ESA created Industry Summits at each conference with a panel of experts focusing on issues and challenges unique to their industry. There were specific summits for locators, gas and oil, telecommunication, excavators, water/sewer, electric, and One-Call.

These are many of the initiatives Scott and his team started, led, and shaped, but they only scratch the surface of what is generated within the industry. Each stakeholder and organization has immense potential and creativity to make the profession safer and better. In September 2025, Scott was honored by the Facility Notification Centers Association at their 50th Anniversary celebration and reunion. His award was a "recognition of your dedicated service, leadership, and commitment to the One Call community." While many of the initiatives he undertook were focused on 811 Center activities, many were aimed at other stakeholder groups. And while ESA and its conference ended in 2024, this book serves as Scott's final opportunity to bring those voices together. Thank you for joining this conference in a book!

In the chapters ahead, we highlight dozens of other initiatives and efforts. This short chapter serves as a guide to some of those initiatives and simplifies some of the names. But it also demonstrates how active the damage prevention community truly is. The activity within this industry directly relates to the safety and effectiveness of what they do, and by extension, how everyone outside the industry lives their daily lives.

Damage prevention professionals have an enormous impact on the nation and its people, even when they don't realize it. In fact, it is the very success of this industry that keeps people unaware – the bombs that never go off. But once someone becomes aware of the industry, and peers into all the activity happening behind the scenes, they can never overlook it again. The chapter ahead gives you that look under the hood. Soon, you'll see 811 everywhere. Just as once you notice the spray paint, it's all you'll see any more wherever you go.

Chapter 7 Resources

ACTS Now – From the Publisher: What's in a Name?

Town Hall Library

Pipeline Ag Safety Alliance

Global 811 Magazine

Excavation Safety Guide

Global Damage Prevention Summit and State Summits

Note

1 CGA awards. CGA – Common Ground Alliance. https://commongroundalliance.com/Membership-Engagement/Awards

Chapter 8

Awareness, Education, and Culture

The rush out the door was her typical frantic chaos – *had she watered that plant, where were the keys again, did she check in for her flight?* Ordinarily, Cynthia was as organized as they come, but travel days always threw her for a loop.

Hands full of keys, the handle of a go-cup of coffee, and a breakfast bar; her purse haphazardly swinging from her elbow; and her suitcase swerving anything but dutifully as it dragged behind her, Cynthia finally made it out the door.

There was no real reason for rushing. She lived less than 30 minutes from the airport and she had in fact already checked in 24 hours ahead of her flight, which by the way did not depart for another three hours. Once in the car, this all seemed to resonate with her and her breathing settled as she said a final prayer for safe travel while loosely holding to the steering wheel.

Opening her eyes, it was a new Cynthia. Composed, calm, and with a rising excitement about the trip that felt more like determination than mere energy. This was her first conference, and she could not wait to meet people and learn.

As she rolled out of her neighborhood, she could see there was spray paint all along the entrance. The city had expanded the sidewalk in an adjoining neighborhood weeks earlier and was in the early phase of doing the same in hers. It may be her first conference, but she was certainly not new to the industry. Cynthia had worked at a call center for nearly six years before moving into an administration and planning role at a major construction company. Needless to say, the spray paint on the street and sidewalks was not lost on her. She may have even noticed it over and above the stop sign she had to heed to get out onto the main road.

It was just before 9 am and traffic was doing its thing, so she was not alone on the road. Like a neon sign, she noticed every shovel-logoed 811 bumper sticker affixed to utility vehicles as she went along. She even noticed two billboards advertising the importance of calling 811 before digging.

At the airport, she parked in a covered garage. More artfully than when she left the house, she looped her purse over the handle on her roller bag and calmly walked down a breezeway to the security terminal. Inside was not very busy, but it still took 15 minutes to zigzag through the rope lines and make it past the screening.

Hopping on one foot as she got her second shoe on, she proceeded to her gate. Having left her go cup in the car, she stopped at a Starbucks near her gate and finally sat down with more than an hour still to go – in fact the board was still displaying the information for the flight before hers. But this was heaven to her – to be on schedule, settled, and sipping fresh coffee.

Sitting there with time remaining also meant an opportunity to socialize with surrounding passengers. She loved to talk about her work, because no one seems to know about damage prevention even though everyone relies on uninterrupted services all day long.

DOI: 10.1201/9781003796619-10

Her first bench partner was not looking for socializing and promptly inserted headphones. But she hit her stride with the next. They talked for a while before the other woman's gate was changed and she kindly departed, wishing Cynthia well and safe travels.

When it came time to board, Cynthia made it to her seat and waited to see who would join her. Only a few minutes in and she had a seatmate. A doctor.

They were very similar and each of them felt at ease letting conversation flow. The doctor was a woman named Martha, about ten years older than Cynthia, and they came from similar backgrounds. When Martha asked Cynthia where she was headed, she could see the light in her eyes before she even responded.

Cynthia shared that she was headed to a damage prevention conference and would get to see the technology expo and meet lots of new people. She was especially excited for the sessions and roundtable discussions. Perhaps unexpectedly for Cynthia, Martha was very engaged. She was familiar with 811 although had never used it. But then she told Cynthia something that really shocked her.

Once, an excavation project outside of her hospital cut a line. Of course, the hospital had all sorts of backups and failsafes, including basement generators for power and multiple hardline fiber lines for internet. But this was one Cynthia had never heard about – an oxygen line. All of a sudden, the stakes seemed much higher to Cynthia.

Of course, she had heard the horror stories of natural gas explosions and the misfortune of sewage leaks, but somehow a medical gas pipeline system seemed to top them all. They spoke for the full two-hour duration of the flight, swapping stories from their respective industries. They found many points of overlap, and both thoroughly enjoyed learning from the other. Cynthia even collected a few anecdotes she would be able to share at the conference – after all, education and awareness are at the heart of why she was attending and why the conference was put on in the first place!

The damage prevention industry we know today really took its shape around 30 years ago. While 811 Centers trace their origin much further back in the 20th century, it was still a local and regional phenomenon. From the 1960s through the 1980s, local utilities and cooperatives were coming together, forming call centers, and formalizing best practices for their own locales. By the 1990s, a unified focus on damage prevention began to take shape at a national level. This, in turn, began to turn "damage prevention" into a nationwide industry, even while it still focused on protecting local facilities all across the country.

Six prominent pipeline accidents between 1993 and 1994 would set in motion a series of events that ultimately birthed the modern damage prevention movement, including the Common Ground Alliance (CGA).

Illustrating how nationwide the issue was, these six occurred in Pennsylvania, New Jersey, Wyoming, Minnesota, and Virginia. These consequential half dozen incidents prompted the National Transportation Safety Board to convene a workshop to bring stakeholders nationwide to one room to elevate key root causes and learnings.[1] The 1994 workshop was followed by reports and recommendations by the NTSB in 1997. With momentum building, the USDOT undertook the *Common Ground Study*, which was published in 1999. By this point, a single, national damage prevention language was truly born.

Prior to 2000, damage prevention education and communication among the different stakeholder groups were done at the local level, primarily through 811 Centers (admittedly still known as "One-Call Centers"). While many of the most active centers proactively shared information and ideas among themselves through OCSI, there was no formal mechanism that brought all stakeholders together. In 2000, CGA was formally established, providing a single roof to house all the nation's stakeholders, and it became the leading member organization. Producing not only damage estimates but best practice guides and eventually white papers and more, CGA put awareness and education center stage. If before, damage prevention was a local concept only active stakeholders knew about, the advent of CGA marked a nationalization and unification of messaging that broadcast damage prevention education and resources from the mountain top.

Common Ground Alliance Primary Stakeholder Groups

The CGA consists of 16 primary stakeholder groups, each representing a different industry or sector involved in the underground utility industry. These groups include:

- Electric
- Gas Distribution
- Railroad
- Engineering/Design
- Insurance
- Road Builder
- Equipment Manufacturing
- Locator
- State Regulator
- Excavator
- One Call Center
- Emergency Services
- Gas Transmission
- Oil
- Telecommunications
- Public Works

One of the first tasks the CGA took on was to develop a Best Practices Guide for the damage prevention industry. Each practice must be approved via unanimous consensus by representatives from all 16 stakeholder groups. The CGA did not waste any time and released the first edition of the Best Practices Guide in 2001, the year after they were founded. Writing these in a form that got all 16 stakeholder groups to agree was a major feat.

As the CGA evolved and grew, it became the central depository for ideas and programs developed at the grass roots level, and its Educational Programs and Marketing Committee also created its own campaigns and educational materials. This organically mimicked the way damage prevention has virtually always been – with both bottom up and top down influence – but this time having a cohesive player in the middle to organize and execute for maximum effectiveness. The CGA often took the grassroots ideas and campaigns created by 811 Centers or other stakeholders and made the ideas available nationally. Many of these local programs were turned into "playbooks" for any stakeholder to use, making it easy for stakeholders all across the United States and Canada to implement the idea in their area.

One of the strengths of the CGA is that it is member driven, and all the committees are composed of its members. This ensures all stakeholder groups and their interests are represented in any and all official actions of the organization. The members are all volunteers, and their time and travel costs are paid for by their employers. This shows the commitment level of the people and companies who take an active role in damage prevention. All of these committed stakeholders also pay member dues to sponsor and uphold the work CGA does. CGA staff provided the organization and support to keep the committees going, and it still does today.

This member-driven format can be very effective, but as all volunteers, there is no way to ensure equal engagement from all 16 stakeholder groups. All stakeholder groups are welcome and encouraged to join in, but a number of the stakeholder groups do not participate on these committees. During Scott's 10 years on the Education Committee, everyone worked hard to be sure that all programs considered every stakeholder group's perspective, but it was difficult without representatives from the group participating. Some stakeholder groups have a hard time justifying the cost of participation, but with today's virtual options, that barrier is gone.

It is worth noting that much of the awareness and education CGA helps facilitate is outward toward the public to improve knowledge of the 811 system and reduce no-notification damage. But another portion of their efforts is aimed squarely at the industry itself to help improve knowledge of those very awareness programs and help the industry refine their practices for better safety and efficiency. One of the intra-industry efforts is awareness and education toward utilizing the DIRT platform. Along with this, an interesting paradox can occur – that with more awareness comes more participation in reporting damage. The awareness efforts can help drive excavation damages down while simultaneously driving up reporting to DIRT. This can give the impression of damages getting worse when in reality it is better data collection.

Put more succinctly by Mike Sullivan,

> Essentially, the more awareness various groups like the FNCA, CGA, CCGA, ACTs Now, not to mention the countless utility companies promoting awareness of their buried plant, etc. create of the existence of DIRT, buried utilities and how to work safely near them, the more damages will be voluntarily reported, which in turn can create the illusion that the situation is worsening rather than improving. To the untrained eye or casual observer, the results are counterintuitive; ie: we appear ineffective as a damage prevention industry when in fact, we are making vast improvements from one year to the next.

The line between public awareness and industry awareness is blurry because ultimately even utility companies are the public. The message of notifying before every dig is for everyone, even if you're employed by a utility company or are a locator yourself.

Instrumental in working alongside PHMSA and the FCC to designate 811 as the nationwide call before you dig number, CGA created a language and vocabulary we still use today. In 2006, after 811 was earmarked and set to go into effect in May 2007, the CGA Educational Programs and Marketing Committee created a special 811 Task Team to tackle the massive task of creating and launching the 811 number and brand. The team is featured in the photo below, receiving the CGA Ron Olitsky Award from Bob Kipp (CGA President) and Erika (Andreason) Lee (CGA Executive Vice President) along with CGA staff.

Having the 811 phone number assigned was a first step, but there remained a huge amount of work to turn it into a program. This team worked tirelessly for over a year to create an 811 logo, tag lines, and a launch plan.

In 2008, CGA was even honored alongside National Launch Partners including PHMSA, with a prestigious Silver Anvil Award from the Public Relations Society of America. This award, bestowed in Time Square in New York City, validated the "8-1-1 Call Before You Dig campaign" for "excellence in influencing public opinion" and CGA as one among many "talented, resourceful, and ingenious organizations that execute strategic public relations planning and implementation."[2] The award came just after the one-year anniversary of 811's full nationwide implementation, demonstrating that the initial awareness campaign made its strongest impact right out of the gate.

Going forward, CGA has served as a facilitator of many key campaigns. While CGA does not fund media campaigns itself, they do create tools, which can be downloaded, built off of, or otherwise used by other stakeholders. This process is a big plus for the industry, because it helps keep the messaging consistent and because many smaller stakeholders could not afford to

Kevin Chumura (JULIE), Tom Shimon (Kansas 811), Linda Sims (Consumers Energy), Gina Greenslate (Panhandle Energy), Patti Lama (Portland General Electric), Bob Kipp (CGA), Dan Meiners (Underground Safety Alliance), Erika Lee (née Andreasen) (CGA).

create these tools themselves. A few great examples of the tools the CGA has funded and their committees have created are:

- The 811 logo development and launch plan
- The Pirate Video: A fun, pirate-themed 7-minute video aimed at children ages 8–11, teaching the importance of calling 811.
- The 811 Agriculture Video: A testimonial-style video featuring an agricultural dig-in scenario, available in English and Spanish.
- Safe Digging Toolbox (5 Steps to Safer Digging): A straightforward overview of the five critical steps for safer digging. Available in both English and Spanish, along with support materials.
- Numerous Public Service Announcements (PSAs) in multi-media formats

On occasion, CGA does manage campaigns, which may include soliciting voluntary funds from their member companies. Probably the most visible example of this effort has been their campaigns to get horse racing jockeys to wear the 811 logo. In 2014 and 2015, jockey Victor Espinoza wore 811 when he won the Kentucky Derby in back-to-back years, exposing 811 to millions of viewers.

Now, there is ubiquity to this messaging. Everywhere from utility vehicle bumper stickers to road-side billboards, "Call Before You Dig" and "811" are everywhere. CGA research demonstrates that awareness is continuing to grow every year, with fewer people having not heard about 811 as time goes on. Even beyond horse racing, NASCAR fans and others see the messaging in branding and sponsorships that continue to seep into every corner of the popular consciousness.

It has all been so successful that in some cases, it is a victim of its own success! In effect, "811" has become a name instead of a number. Scott has a friend who once asked "*what is the number for 811?*" He knew about calling before you dig after being around Scott for years, but 811 being a phone number didn't register. And while that's a true story, it closely parallels a familiar joke "what is the number for 911" popularized in Little Rascals and The Simpsons. We can joke about it only because it is so simple and so universal. It underscores the importance of how effective awareness must move beyond simple familiarity but into genuine education.

Public Awareness Versus Damage Prevention Education

The U.S.'s PHMSA and Canada's CER regulations mandate that pipeline operators attempt to educate contractors, emergency responders, public officials, as well as property owners and school officials near their pipelines. This typically involves extensive mailings, training meetings for emergency responders, and educational meetings for contractors and school officials. The term PHMSA, the CER, and the industry use for all this education and communication is "public awareness."

One element of public awareness is marking signs, which we covered in Chapter 6. While historically, all stakeholders did this, only pipelines are required to by federal rules. As the images in that chapter demonstrated, sometimes these signs can become dated, but increasingly, modern technology is future-proofing them. As Mike Sullivan explains,

> Marker signs are a last line of defense and should be maintained. And today, in addition to a phone number or 811, we have the ability to affix a QR code to direct an excavator to ClickBeforeYouDig.com where they can submit a locate request to their respective center.

Many companies – primarily pipeline operators and gas distribution companies – have separate departments for public awareness and damage prevention. On the one hand, one can understand this, because they have very specific public awareness mandates from PHMSA and the CER, but on the other hand, how can educating contractors, land owners, public officials, emergency responders, and schools about safe excavation and pipeline safety *not* be considered damage prevention? Having everyone involved in any type of safe excavation education all working together seems like an effective approach. For our purposes when we refer to damage prevention we are also lumping in public awareness because, as this entire book seeks to emphasize, damage prevention is multifaceted and has to involve not only different stakeholder groups and their perspectives but different approaches to the same problem. In effect, awareness relies on education, and education relies on a pervasive culture.

A final note that applies to both federally required public awareness and industry-led damage prevention communications is a critical dynamic – the two-way nature of effective communication. Like many aspects of damage prevention, when multiple parties are communicating or collaborating, each can learn from one another.

Companies that do good public awareness are not just pointing a megaphone outward, but are listening to the feedback that comes in. This helps them hone their messaging, identify challenges the public are facing, and more.

This emphasizes the importance of listening as an underlying necessity to good communication. As George Bernard Shaw once said, "The single biggest problem in communication is the illusion that it has taken place." But to ensure public awareness and robust communication does in fact take place, awareness cannot simply be about sending out resources, it must include taking in the responses. That takes a leadership class in tune with the goals and purpose of public awareness and damage prevention. In turn, that relies on a deep-seated commitment and culture.

Culture from the Top

One critical aspect of culture that cannot be overstated is that if you want the impact to be felt in the field it must start at the top. While we often think of culture (rightly) as an organization-wide or industry-wide concept, it must be present even with a few individuals at a work site. Very often, young guys on a site will look up to the veterans on a crew. They'll follow their lead, look for validation, seek to emulate them, or look to prove their value by demonstrating their work ethic. It is essential that the older members of the crew be steeped in the culture of damage prevention and dedicated to following all the rules and implementing best practices. They can also reinforce the mentality that what they are doing on the site is a profession, that they are skilled and important, and that even young guys and gals simply breaking up concrete or moving dirt are participating in something bigger than themselves and have a role in something successful.

The most significant impact of culture on safety is at boots-on-the-ground level, but a company-wide culture of damage prevention must start at the top, which means the C-Suite or the owner. If executives at the top talk about the importance of damage prevention and excavation safety, that is great, but taking actions that support and demonstrate the verbal commitment is what creates the culture. Doing this consistently over the long haul is what will impact the real people in those boots on the ground.

Wylie Davidson is a top-notch speaker on safety culture. He has spoken at ESA conferences multiple times, and he always has a packed session with extremely high ratings. In a recent 2025 LinkedIn post, Wylie shared an observation that serves as a salient example of how culture starts at the top. He wrote about a conversation with the owner of a tree-trimming service that was doing work in his neighborhood. As Wylie watched them work on a tree in his front yard, he couldn't help but notice the lack of safety gear employed by the trimmer and his team.

Wylie gets an *A+* for taking the initiative to chat with him about the lack of safety gear. The owner's response was "they don't listen to me, I told them to wear it." Not surprisingly, Wylie points out that the owner himself wasn't wearing any PPE! At one point, Wylie even offered to give him earplugs and glasses for his crew, but he refused. "They won't wear it," the owner deflected.

As Wylie aptly points out in his post, "This was a classic example of what not to do. Starting with leadership. Are your team seriously gonna do what you ask them to do if you don't do it yourself?" He closes his message with a perfect summary: "Safety starts at the top, with leadership. Don't excuse it away, demand it."

Of course, creating culture can be difficult. That is why borrowing and learning from others is an important aspect of leading. It also emphasizes and shapes the culture to be one of constant learning, humility, and collaboration. In the damage prevention industry, there are many organizations generating educational, awareness, and other resources that can even help inform business owners on ways to implement training and best practices to reinforce culture, prevent damage, and keep excavators safe.

National Organizations with a Damage Prevention Mission Make Everyone Safer

The CGA is a great nationwide organization that has members from all stakeholder groups, but there are many additional stakeholder groups with a more narrow focus that make a positive impact on safe excavation. Admittedly an incomplete list, the organizations mentioned here are enough to gain a glimpse into the dynamic and highly active industry helping churn out the culture of education and awareness. We encourage readers to look into these organizations and the many others out there doing this important work.

The Pipeline Association for Public Awareness (PAPA) is a group of pipeline operators who teamed up to work collectively to meet PHMSA's public awareness requirements, and more importantly, to promote pipeline safety. As a group, they are far more effective and efficient with their mission because they can pool resources to reach more people while avoiding some of the repetition and message clutter that took place when many of their individual efforts were overlapping.

The North American Telecommunications Damage Prevention Council (NTDPC) is a group of major telecommunications companies that compete in a number of areas, but realize that together, they can have a positive impact on public safety. Their stated goal is "to prevent damage to the buried facilities that form the telecommunications infrastructure of the United States and Canada. End-to-end integrity and the ongoing reliability of these facilities are vital to our shared national interests."

Facility Notification Centers Association (FNCA) is an industry association made up of the vast majority of 811 centers across the United States and Canada. The objective of FNCA is to evolve the industry by facilitating collaboration among one-call notification centers. Most grass roots education ideas come from 811 centers, so FNCA provides an ideal vehicle for sharing these ideas. Many of these successful programs are passed on to the CGA, which shares them with members that may not be a part of FNCA.

The Distribution Public Awareness Council (DPAC) is a consortium of natural gas utilities dedicated to addressing the unique challenges of public awareness in natural gas distribution. As of September 2025, DPAC represents 31 states, more than 500,000 miles of distribution pipe, and serves over 43 million natural gas customers. DPAC fosters collaboration and sharing best practices, while working to improve community safety nationwide.

Using the well-known "stop, drop, and roll" fire safety program as a springboard, DPAC created a "detect, dash, dial" campaign to create a simple and memorable message to increase the general public's awareness about how to recognize and respond to a natural gas leak.

DPAC developed this public awareness program to ensure people know what to do if they smell gas.

The Drain Tile Safety Coalition exists to bring together best practices in safe drain tile installation. Drain tile is a plastic pipe with holes in it so water can enter the pipeline and be carried to ditches, streams, rivers, or a managed drainage system. This allows farmers to improve productivity and can also protect building foundations. Scores of farmers safely install tile and reap the benefits every year. Unfortunately, a handful roll the dice and start drain tiling projects without first having underground pipelines and utilities marked. While it is the exception and not the rule, the Drain Tile Safety Coalition shares examples of past incidents on their website to demonstrate how accidents happen and the consequences of hitting a line. As the coalition points out to its audience, these accidents could have been avoided by requesting a free locate to have lines marked. One example from their website that is representative of the outreach is the video *Three Seconds Later*. Here is a brief video description along with a link to the 12-minute video:

> On an early December morning in 2017, two farm workers and two farm owners were installing drainage tile on a piece of rented land in Illinois, land they had farmed for decades. One of the farms' owners was using a tractor to pull a tiling plow. The tractor pulling the plow became stuck. In the process of trying to free the tractor pulling the tiling plow, the workers hit a 20-inch natural gas transmission pipeline. The escaping natural gas ignited almost immediately, killing the two owners and injuring the two workers. The fireball could be seen for miles. The fire destroyed two tractors and two trucks and burned for hours.

The Pipeline Ag Safety Alliance (PASA) was established in 2015 as a collaborative public awareness and damage prevention program designed to enhance and spread utility safety awareness among agricultural professionals nationwide. PASA uses an "educate the educator" concept through a partnership with the National Association of County Agricultural Agents (NACAA). PASA works directly with extension professionals educating them on safe digging practices. These agents have a significant reach into rural communities, capitalizing on the trusted reputation that agricultural agents have cultivated with farmers and ranchers in their respective counties. PASA reaches a large segment of the ag community with actual education versus simply ads or mailings – a key data-based distinction between awareness aimed at basic familiarity and education aimed at behavior modification.

Coastal and Marine Operators (CAMO) is an organization made up of stakeholders with assets underwater. This underappreciated aspect of damage prevention is critical for pipelines, fiber, and more that lie below the surface of the water, where non-traditional excavation risks exist like dredging, pile driving, and anchor dragging. Through awareness, training, and collaboration with industry and agencies, CAMO advances reforms that strengthen environmental protection and infrastructure integrity. They are helping prevent some of the hardest incidents to stop and clean up, and have also formed the new MarineSafe811 initiative to improve locate requests in marine environments and provide operators with more tools to safeguard their facilities. CAMO collaborates with the 811 Centers in Texas, Louisiana, Mississippi, Alabama, and Florida, with more likely to join in the future.

The Cross Bore Safety Association (CBSA) is an association of volunteers who have a mission to prevent cross-bores. In 2019, CBSA released the *Leading Practices for Cross Bore Risk Reduction*. This 88-page document provides suggested procedures and steps that can be taken to reduce the risk of cross-bores.

The Pipeline Safety Trust (PST) is dedicated to more than just damage prevention, but overall pipeline safety. Their work involved research and policy reform efforts among many other initiatives to help prevent future tragedies. They were founded as a direct result of a major excavation damage incident.

The *Communications Infrastructure Contractors Association (NATE)* whose members build the 200 to 400-foot towers that keep us all connected to our digital lifelines is extremely focused on safety. This emphasis extends beyond the obvious fall protection into excavation safety, because all cell towers are connected to both electric power and the communications network. This translates into excavation being a part of every tower build. In October 2025, NATE unveiled a professional training video titled, "811 – Call Before You Dig" as a part of their Climber Connection video series.

The video presents a compelling narrative from the perspective of a tower technician whose work could place them at risk if buried utility lines are struck. Through a blend of real-world footage, animations, and expert commentary, the video illustrates how even shallow digging or ground disturbance can interfere with concealed utilities. The informative video also emphasizes person-to-person responsibility and highlights how contractors, technicians, and site supervisors all share accountabilities. Additionally, the video serves to reinforce the steps involved in the 811 process, such as how to submit locates, what to expect, and best practices when waiting for clearance; and highlights the need for proactive communication on every job site, before any shovel or machinery touches the earth.

Alongside these organizations and groups, many other apps, programs, and initiatives add another layer of impact by reaching individual excavators and homeowners directly. Many more organizations exist or will exist in the years ahead. This is all a benefit to the industry and to the public. Regardless of how specialized a part of the industry is, it is vital to tell the same story – that culture must be informed by reality, best practices must be employed by all stakeholders, and effective awareness must include education beyond only slogans (which admittedly are an important part of memory!).

Taking these effective elements – whether local or specialized – and amplifying them widely saves resources, ensures accuracy, and maximizes effectiveness.

National Campaigns Create a Focus and Garner Media Attention

The CGA has organized two recurring annual campaigns that have been tremendously successful in terms of getting all stakeholders involved and in gaining media attention. The CGA launched National Safe Digging Month (NSDM) in April of 2008 as a nationwide awareness initiative to remind homeowners and contractors to call 811 before any digging project. April was chosen as the start of the spring digging season in much of the country, when outdoor projects like landscaping, fencing, and construction pick up dramatically. The NSDM concept spread to Canada, which likewise promotes April as Dig Safe Month.

The CGA provides a comprehensive tool kit including press release templates, graphics, Earth/Arbor Day campaigns, Governor's Proclamation tools, audio files for radio ads, PSAs, and more. Providing these professional, ready-to-go tools makes it incredibly simple for any stakeholder, regardless of size, to make a NSDM push, and it helps create consistent messaging. Even with all these awesome templates, there are still many unique grass roots promotions that pop up during NSDM.

For those new to the industry, you may not have connected the fact that 811 can also be associated with August 11th, which is known as 8/11 Day. The success of NSDM motivated the CGA to look for another reason to get media coverage and energize stakeholders to promote safe digging. Promoting August 11th as 811 Day was a natural fit.

As with NSDM, the CGA provides a tool kit in order to make it easy for stakeholders to participate with a consistent message. The biggest difference is that all the 811 Centers and stakeholders efforts are focused on a single day rather than being spread out over 30 days. This creates an extremely intense effort, which permeates the media – especially social media.

Kudos to the CGA for creating and consistently supporting NSDM and 811 Day. The 811 Centers get very energized by these programs and turn them into incredibly effective public awareness tools. These two campaigns often get stakeholders who may not normally promote 811 and safe digging, in terms of public awareness, to amp up their efforts. Both events also spawn many new grass roots ideas which may end up being used repeatedly by many other stakeholders. Two additional efforts highlight that effect.

As we covered in the previous chapter, ESA launched Locator Safety & Appreciation Week (LSAW) in 2015. LSAW takes place during the last full week of April (fittingly within NSDM) as a way to highlight a specific stakeholder group within the overall safe digging process, LSAW provides an opportunity for the industry to thank our boots-on-the-ground heroes and let them know their contributions to safe worksites and communities are valued. The intent is to provide tools so companies of any size, in any industry, can participate in LSAW. ESA created a logo, posters, appreciation card graphics, social media ideas, along with a list of simple internal steps a company can take.

This program is aimed at recognizing the efforts of all locate technicians.

Modern Roots March

Building off of the idea of safe digging month, the Alliance for Innovation and Infrastructure (Aii) looked for ways to increase the effectiveness and reach of the damage prevention industry's message. While not a stakeholder group – Aii is a public policy think tank – Aii settled on a complementary campaign that would help amplify NSDM before it arrived.

Starting in 2020, Aii began leading the annual national *Modern Roots March* campaign. Held the month before April's NSDM, it builds public intuition about why calling before digging is essential by drawing attention to the infrastructure beneath our feet. Just as everyone knows a tree has roots, *Modern Roots March* encourages people to recognize that every element of modern life – from buildings to broadband – has "roots" in buried pipes, cables, and wires.

Serving as a primer for NSDM, the campaign broadens public understanding of the systems that sustain daily life while promoting awareness, damage prevention research, expert insights, and safe excavation practices.

Without the member-driven work of CGA, there would be less complementing efforts and nationwide awareness in the damage prevention industry. While much of this awareness and resource sharing takes place virtually, there are elements of education and awareness – and definitely culture – that cannot be passed along in an email or social media post. Some things require rubbing elbows.

Face-to-Face Sharing of Ideas Is the Most Effective Damage Prevention Tool

Communication has always been the cornerstone of damage prevention, and the best communication happens in face-to-face settings, where both learning and issue processing can take place. The 811 Centers organized the annual International One Call Symposium beginning in the late

Free, shareable poster graphics helping build intuition about underground infrastructure.

1970s. This conference covered operations as well as education and damage prevention, but it was primarily attended by only 811 Center employees and their Board members.

Scott felt that for damage prevention to become a unified effort, and to solidify its status as a standalone industry, there needed to be a national conference and trade show for everyone involved in all aspects of damage prevention. He had a good friend, Greg Geisler, who was the co-owner of a successful trade show company, so he approached Greg about starting a conference/trade show for damage prevention. Greg liked the idea, so Scott suggested that Ron Rosencrans, founder and publisher of Underground Focus Magazine, team up with Greg to put together the education portion of the event.

Greg's company, Champion Productions, launched the Damage Prevention Convention (DPC) in 1997, with Underground Focus creating the education and Scott acting as an industry consultant. The DPC became the place for the industry to come together to exchange ideas and discuss problems. As the CGA started to gain traction after being formed in 2000, Scott suggested the CGA hold an annual meeting at the DPC, which they did for several years. Greg sold Champion, and the new owners did not have the same commitment to the industry, so Greg left. Scott saw the change and severed his ties to the DPC as well.

As overviewed in the prior chapter, Scott launched what became ESA to put on a number of conference and other initiatives, the first of which took place in 2004 in Los Angeles. The goal was to combine classroom education, with outdoor demonstrations and a trade show – all focused on damage prevention and safe excavation.

In 2005, CGA decided to take a fresh look at what event might be the best place for them to hold their annual meeting. Over 15 events submitted bids in response to the Request for Proposals (RFP) to host the CGA annual meeting. ESA's event was chosen to begin in 2006, in part because ESA was willing to add "CGA" to the front of the event name.

The CGA held its annual member and committee meetings at the Excavation Safety Conference, but it did not own, finance, or produce the conference or trade show. This arrangement proved ideal – CGA could concentrate on its meetings and member engagement without assuming the financial risk of running an event. ESA financed and produced the conference, developed the sessions and trade show, and managed global marketing.

From 2006 through 2019, the "CGA Excavation Safety Conference" hosted the CGA's annual meetings and grew into the world's leading conference and trade show focused on damage prevention. This was a major win for the industry, providing a central venue where all stakeholders could meet face-to-face, share ideas, and collaborate on solutions.

Following the launch of 811, ESA added "811" to the Excavation Safety Conference name and logo to help build brand recognition and public awareness. The branding impact was immense – during the 12 months leading up to the 2019 Excavation Safety Conference alone, ESA generated more than 63 million impressions of the 811 logo. That level of annual exposure became consistent after "811" was incorporated into the event's name in 2013.

In addition to amplifying 811, ESA's branding efforts also boosted the visibility of the CGA name, which was featured in the event title from 2006 through 2020. Through its magazine, the Excavation Safety Guide, and extensive digital marketing, ESA provided exceptional early exposure and recognition for the CGA brand.

Over the years, the conference grew to nearly 2,000 attendees from around the world. With over 50 sessions and excavator, electric, pipeline, water/sewer, one-call, and fiber summits, this event became the global hub for damage prevention and excavation safety education. Often, over 100 attendees from Canada participated and spoke at the event. After recognizing the benefits of the event, the Canadian CGA (CCGA) decided to launch their own event, the CCGA

Damage Prevention Symposium. Unlike the United States, which was owned and operated by ESA, the CCGA owned their event and hired ESA to produce it.

The largest contingent of overseas attendees at ESA's conference was always the Australians. There was no event in Australia that brought all stakeholders together nationally, like the U.S. event, so in 2018, ESA launched the Oceania Damage Prevention Conference (ODPC). ESA teamed up with PelicanCorp, who was extremely committed to damage prevention and knew the market. The ODPC saw immediate success and has since accelerated global sharing of ideas and best practices.

During the period of COVID-19, ESA began monthly virtual Town Halls to allow all stakeholders to learn and participate in discussions on industry topics. This virtual format attracted attendees from across the globe and allowed stakeholders who did not have travel budgets to voice their opinions and learn from their peers. While in-person meeting has a deeper connection, remote activity skyrocketed, and the increased regularity of monthly town halls helped bolster the dialogue in a conference-like way but with laser-sharp focus on individual topics. These could be fleshed out more at conferences, while presentations made at the conference informed future town hall conversations.

The industry has increasingly utilized multimedia platforms to advance in-person learning and influential engagement. Beyond ACTS (who took over the ESA Town Halls in mid-2024), regular roundtables by Planet Underground and others bring stakeholders into one room for critical dialogue. These are recorded and streamed to ensure the in-person dynamics can be witnessed by even larger audiences. Likewise, podcasts have emerged as a strong conveyer of person-to-person connection with the host and audience, while being disseminated virtually far and wide. Successful podcasts from *The Safety Moment* by Utility Safety Partners to *Coffee with Jim & James (and also Ashley)* not only bring individuals from across stakeholder groups in front of one large and diverse audience, but they often take place at conferences in the exhibit hall or in a common area. Another thing that is often spotted at conferences, and serves to further education or spur on real-time conversation is one of the trade magazines in the industry like the Global 811Magazine and the American Locator, which are dedicated to the hardworking industry professionals dedicated to a culture of education.

In 2021, the CGA launched its own conference dedicated to damage prevention. With two major events competing for the same spring audience – and many companies only able to support one set of travel and registration costs – the market could not sustain both. As a result, ESA's time serving the industry came to a close in 2024. ACTS seamlessly carried the mantle, inheriting the ESA Town Halls, Excavation Safety Magazine (now Global 811 Magazine), Excavation Safety Guide, PASA, and the Excavation Safety Conference (now the Global Damage Prevention Summit). In many ways, ACTS was uniquely positioned to continue these educational initiatives and expand their reach – further advancing the culture of shared learning and safety awareness at the heart of damage prevention.

By moving the Global Damage Prevention Summit to the fall in 2025, ACTS reignited the international exchange of ideas. FNCA capitalized on this momentum by hosting its fall meetings alongside the Summit, including a special 50th Anniversary celebration of "One Call."

This blending of education for all stakeholder groups and FNCA meetings fosters continued communication within the industry.

Smaller Can Be Better, or At Least More Personal

On a smaller scale than a national conference, but very much an in-person event, many organizations put on meetings, trainings, award ceremonies, expos, summits, and more all across the country.

Many of these events and local initiatives are also put on by or in partnership with local utility coordinating committees (UCCs). These are truly the boots on the ground helping support and improve damage prevention efforts by being in touch with and made out of community members.

Some events occur as often as bi-weekly or monthly meetings, while others are annual gatherings. These offer vital educational and networking opportunities, while also building and maintaining energy and enthusiasm for the work. Events such as Virginia's Local Damage Prevention Committee (LDPC) meetings or Louisiana 811's "Diggers Night Out" and others can fulfill many purposes at once.

"Dozer Day" is an annual event put on by NUCA chapters in a number of states. Community members and children get a chance to sit in real heavy equipment (bulldozers, excavators, skid steers, etc.) with professional operators. The event is designed to teach children and families about construction, excavation, and underground utility safety. There is no better way to learn than this type of hands-on experience. Not only does it create an understanding and appreciation of the contractors who operate this equipment, but it helps make a career in the industry look appealing while reinforcing the importance of safety.

Elsewhere, other stakeholders have their own versions of equipment days and local events. Scott participated in DIGIN Midwest's (formerly MUCA) "Day of the Dozers" multiple times. This is entirely separate from NUCA's "Dozer Day." Participants can really feel the enthusiasm and interest of all the attendees. Scott and other volunteers staffed a game in which kids learned what the temporary marking flags stood for and how important they were. The event also raises money for charity. Putting on these events is a huge amount of work done by volunteers and is indicative of the passion people in the industry have for education and safety. Seeing and interacting with other people, equipment, and demonstrations reinforces education for more lasting impact.

We have explained how a single national campaign to speak with one voice can broaden the reach and impact of the message. We also covered how in-person meetings, particularly at conferences, can build rapport and trust. But there is another way these can both be accomplished simultaneously while also rallying and encouraging industry participants themselves to validate their hard work in front of the public: notable partners.

2015 "Day of the Dozers" at DIGIN Midwest. A massive demonstration of construction equipment, where attendees can actually participate. These events provide hands-on education and encourage people to consider equipment operation as a career. Kids activities teach safety in a fun setting.

Ambassadors, Spokesmen, and Notable Partners Build the Damage Prevention and Excavation Safety Message

The industry has been blessed with a number of tremendous brand ambassadors and partners of all kinds, who work hard to spread the 811 and safe excavation message in some powerful and unique ways. These ambassadors – whether official partners or campaign-specific advocates – come from a variety of backgrounds, which help them reach and relate to different audiences. Some famous figures are even featured in commercials or use their personal platforms for further outreach. These include local heroes, household names, industry veterans turned passionate speakers, and more.

Naturally, when you think about promoting 811, you think of using a rodeo clown. Okay, it may not be the first thing you think of, but JJ Harrison is one of the best-known rodeo clowns in the country. JJ grew up in and around rodeo in Washington state. He competed in roping all the way through college, eventually earning his master's degree at Washington State University. After college, he became a middle school teacher, where he learned how to effectively communicate with kids. While teaching, he started working as a rodeo clown on the side to stay connected to rodeo and make some extra money.

JJ is a natural entertainer and gifted off-the-noggin speaker. Not unlike Robin Williams, he can entertain without a script. He regularly engages crowds with little or nothing to go on – so you can imagine how much more he can do when there's a clear message and powerful story. JJ's connection to 811 began when PAPA became a rodeo sponsor. Truly focused on awareness, PAPA did not want their name promoted, they wanted 811 to be the message. That meant using his relational capital and rapport with the crowd to speak directly to the life and death importance of calling 811. It may be dangerous to share an arena with a bull, but at least you can see it coming, digging without calling 811 would be like putting on a blindfold and opening arena gates without knowing which ones had an angry bull behind it!

JJ loves his connection with 811 and promotes the message virtually 24/7. As JJ points out, rodeo clowns get more exposure than any other participant at a rodeo, and they are the ones connecting with the crowd. This makes them a unique focal point and unofficial master of ceremonies that persists across many events. Even though JJ has other sponsors, he has become known as the 811 Rodeo Clown, which is a great thing for 811. There are even 811 Rodeo Clown Halloween costumes!

During the course of a season, JJ does around 38 rodeos with crowds up to 20,000 or more. He has 150,000 followers on social media. In addition to reaching children (who can learn this vital message at a formative age and never forget it), the rodeo demographic overlaps heavily with home owners, farmers, and contractors who are all likely to dig. When you add in his viral videos on the Cowboy Channel, and other events like demolition derby and speaking engagements, JJ's total annual reach is in the millions. Even more important than the raw numbers is the way JJ promotes 811 with a combination of humor, fun, and passion that somehow gets the safe digging message to stick while making people smile.

JJ reaches new people every year, and even if many people don't know his name, they recognize or have come across him before. But for others – who certainly paid their dues over the long haul – their names already have household recognition.

Mike Rowe is perhaps best known for hosting the Discovery Channel television show, *Dirty Jobs,* which ran for 10 seasons. He is also known for being a dedicated advocate for skilled trades, hard work, and many underappreciated or downright thankless jobs. His mikeroweWORKS Foundation embodies this mission and makes a tangible difference across the country. Supporting over 21 skilled trades and facilitating tens of millions of dollars in work ethic scholarships, it is obvious that Mike's work is truly impacting the future of American industries from the ground up. This positioned him as perhaps one of the singularly best-fitting public figures to help expand awareness of and influence in the damage prevention industry.

Starting in 2017, leading stakeholders orchestrated a campaign featuring Mike Rowe (and an animated counterpart, "micro Mike Rowe") to boost awareness of the 811 "call before you dig" message.[3] Mike Rowe's participation helped the message resonate with the public through engaging educational content that spanned from site visits to encourage workers and hear their stories to video PSAs featuring both live and animated activity. In fact, the micro Mike Rowe character was often helping demonstrate what happens when things go wrong to reinforce the message the real Mike was telling stakeholders and the public. These all had the same aim: to educate the public on protecting buried infrastructure like pipelines, wires, and cables to keep people safe.

Mike's participation really embodied the type of effective communication that goes beyond standard public awareness. He didn't just repeat lines, he listened and engaged with stakeholders for months as the campaign was generated, reinforcing the importance of two-way communication. His partnership across multiple years included radio ads, television, digital, and more. In 2020, he partnered with Iowa One Call – a state he had been intimately familiar with from three *dirty jobs*. They said it was so successful they were actively in talks with his team to do future campaigns.[4]

Many industry innovators worked hard to plan the engagements with Mike Rowe, and tagging him in was a natural fit. By 2022, *micro Mike Rowe* made another appearance that achieved incredible reach.[5] This campaign continued to emphasize the importance of calling 811 before digging near natural gas pipelines. The campaign generated over 422 million impressions, which earned Southern Gas Company the Southern Gas Association Corporate Communications Award.[6]

Mike Rowe was a brilliant choice to help partner with 811 programming, because so many people doing excavation can relate to Mike and his genuine down-to-earth personality. His career has not been lip service, but defined by putting himself in the very shoes of the people doing hard work. That has earned him rapport and credibility that no one can short cut or pay for. An industry partner people trust, relate to, and like, is someone who can not only carry the message farther and to new audiences, but can pat the industry on the back in the process and ensure that as new people learn about it, they hear about the hardworking men and women serving them day in and day out that they never realized.

Many more ambassadors and partners are spreading the safe digging message around. From notable auctioneer Tom DiNardo to prolific actor and voice actor John Ratzenberger, whose work in Cheers, Toy Story, Cars, and other Pixar hits has made him a household voice if not name, to Roger Cook from the home improvement series This Old House, ambassadors and spokesmen come in and out of the 811 orbit regularly to help add new energy into the message and leverage their skills and notability for a noble cause. Professional sports teams regularly link arms with 811 Centers, while athletic legends like longtime sports commentator Bob Costas even joined a 2025 summit as a featured guest speaker.

Every day, more people realize their lives are impacted and become avid spokespeople – in fact, one story in Chapter 13 will highlight just how an average Joe (or *Cliff* to be more precise) went from an apprentice on a job site to a global spotlight, all because of an incident he endured. This helps demonstrate that passionate speakers, innovative ideas, and culture-stoking initiatives can come from anywhere!

811 Centers Create Their Own Ambassadors

In Georgia, the bulldog rules the state. It's as recognizable as just about anything, and it stands for strength and, of course, for football. But there's another dog hard at work in the Peach State helping spread the message of safety: Digger Dog.

While the industry may identify partners and brand ambassadors for the public at large, 811 Centers are often busy creating entirely new ambassadors to reach the next generation. That's where Digger Dog comes in – a mascot designed to introduce the damage prevention industry to younger Georgians starting in elementary school. Not unlike McGruff the Crime Dog, sometimes an amiable animal can help make serious or complicated topics simple.

Keeping things canine, South Carolina 811 has its own *pup* (for those not in the know, this call center used to be known as PUPS, the Palmetto Utility Protection Service). The former PUPS pup is known as Diggity Dog.

We mentioned briefly that personas are a common marketing tool in the industry. From JULIE to Miss Dig, Miss Utility, and more, using names or pneumonic devices has long been an effective strategy. In Kentucky, the program was briefly known as "BUD" – standing for "before u dig" and having a casual but friendly appeal to it. Mascots are an integral part of this and even harken back to the very first one-call program in the country – Mac Mole.

These build on intuition, often involving animals or other personified digging-related concepts. At the USDOT, Hazardous Matt is an animated figure designed to remind travelers and shippers about hazardous materials. Many are simple mascots that serve as an entry point for students, while others have more backstory and development to them, like the Utility Defenders at Miss Utility. Somewhat reminiscent of Captain Planet and the Planeteers or even the Avengers, the Utility Defenders are a group of superhero children both protecting and educating about critical infrastructure. They include members that "each represent a different underground facility: Vapora (gas/oil), Bio-Boy (sewer/drains), Zap (electric), Marina (drinking water), Signal (internet/phone), and Mixy (reclaimed water)."

Animated videos, coloring book designs, stickers, and more help facilitate youth education. From Holy Moley and Gus the Gopher serving as intuitive digging-related characters to Safety Sam and others helping to educate about the best practices, mascots, and statewide educational initiatives spread the message in a way that sticks and reaches many generations at once.

Mascots are not the only resource 811 Centers deploy, as they often partner with local community leaders, former athletes, media personalities, or important businesses in the area. These help get more people involved and increase the interest in the subject. And 811 Centers know who and what is best in their state or locality to help strike the right chord.

These initiatives and campaigns can even be boosted through other age-appropriate partnerships, like formal partnerships with the Boy Scouts of America. Many 811 Centers have video PSAs with young men or troop leaders sharing the important message about calling before digging. This can take the message and let other organizations with great credibility and reach serve as ambassadors to their own members and audiences so that the critical safety information is propelled outward many times over. Some of these may even inspire others to start their own initiatives.

Grass Root Initiatives Keep New Awareness Ideas Flowing

811 Centers and national organizations create some excellent, effective programs, but the fresh ideas that are sparked by the passion of individual stakeholders or stakeholder companies create a constant flow of innovation. We share only a small sample to provide a look into how these happen organically, but there are new ideas being created all the time. It wouldn't be possible to capture them all here, so let these that only scratch the surface inspire you to create your own or share examples you know with more people.

811 Runs – As a long-time runner, it dawned on Scott that an 811 Run might be a good public awareness event, so he sent an email proposing the idea to Whitney Price, a college runner. Whitney was a Rhino team member at the time and had attended several CGA Committee meetings as well as ESA's annual conference. As soon as she got Scott's email, she walked into his office and the brainstorm began. The idea behind organizing a road race called the 811 Run was to create public awareness of 811 via the marketing as well as at the race itself.

The first race was held in March 2013 in West Palm Beach the day before the Excavation Safety Conference began. Of course, the race was 8.11 kilometers and it started at 8:11 am. Sunshine 811, the 811 Center in Florida, promoted the race aggressively in Florida and ESA promoted it to its attendees and potential attendees.

The race was successful, so Rhino and ESA teamed up along with enlisting some help from the local CGA Regional Partner to hold another 811 Run that year on August 11th (811 Day) in Minnesota. At the Minnesota race, local utilities and contractors had excavation safety displays and children's games to educate the general public on 811. One of the many volunteers helping with the race got Clear Channel Communications to provide multiple full-size billboards on the interstate promoting the race. The 811 Run had now crystallized into a high-visibility public awareness event promoting safe excavation and 811 to thousands.

After the Minnesota race, Whitney created a "playbook" with some other industry professionals who were excited about the potential of the idea. She held several meetings to share what she had learned as well as come up with the best way to put on an 811 Run. This playbook was given to the CGA so anyone interested would have a great place to start. The Minnesota Regional CGA took over the Minnesota 811 Run after a few years. There are now regular 811 runs in at least nine different states. The creation of the 811 Run is a great example of how his industry runs on the passion of the stakeholders. There was no financial gain for Rhino, ESA, Whitney, or any of the volunteers; it was done purely to pursue their passion for saving lives.

The 811 Runs drive home the 811 message during the marketing campaigns and during the events.

811 Mudder – Tough Mudder events are unconventional events that combine running, crazy obstacles, and lots of mud. These events can definitely push you to your physical and mental limits, and yet somehow they are fun. They focus on teamwork not individual performance. You really cannot get through or over some of the obstacles without help from someone. Teamwork and mud should already place the reader in a damage prevention headspace.

The course length and obstacles vary, but they all involve physical challenges, lots of mud, and electric shocks. In 2014, the Rhino and ESA teams got together to enter the Minnesota Tough Mudder as a team building event and to promote 811. The course was about 10 miles long, with 22 obstacles. Participants climbed over walls, slithered through mud on their belly trying to avoid the live wires dangling inches above their back, jumped into an ice bath, and many more. What could be more enticing?

To be sure, 811 was front and center. They called their team *811 Mudders* and created a logo and brightly colored shirts to wear during the event. Since an average of 10,000 people enter these events, it seemed like a great way to share the word on 811. To expand the reach past Rhino and ESA members, they recruited friends to join the team, and Rhino provided free 811 Mudder shirts to wear at the event.

As it turns out it was very effective at promoting 811. Not only did all the team members become 811 ambassadors, but many other participants got the message. Between obstacles you run, jog, or walk to the next obstacle, and thanks to the large 811 on their shirts, everyone on the team was asked what 811 was more than once during the event. Sundar Ganeshan Ayikudi Paramasivan, an ESA Team member, turned this into another playbook and gave it to the CGA, so other interested groups could use it to promote 811 in their area.

These race-related initiatives only further illustrate the point that anyone can create awareness even in the course of doing something fun. It does not take advertising dollars and an official corporate event – sometimes just wearing an interesting t-shirt to an event can be a conversation starter! In fact, an even more organic trend is emerging across social media at the intersection of awareness and education and AI.

Numerous damage prevention professionals from across all stakeholder groups have turned to generative AI (such as ChatGPT image creation) to make visually interesting graphics or posters to draw attention. These often depict common scenarios like someone pulling up colored flags prematurely, sometimes create a representation of the buried

This 811 Mudder flow chart is a good example of how a program can be simplified and passed on to other stakeholders.

landscape to help people visualize what is below, or form idealized images of stakeholders in the field.

This trend is more than social media filler, it's becoming a key way to drive interaction, reach younger people, and communicate essential insights. With an intriguing image to hook readers, many social media users will stick around to read full paragraph captions describing challenge or solutions in the industry. This has equal appeal to those in the industry and helps attract outside audiences to learn more.

When it comes to public awareness and damage prevention, there is no shortage of ideas about how to get the name, brand, and message in front of people. In fact, after the creation of the Safe Digging Partner logo, spearheaded by USA North, anyone can amplify the message and use the logo to drive people toward 811 in a unified and cohesive way. Here are just a few other unique awareness campaigns that sprung up over the years with the originators of the idea:

- An 811 hot air balloon – Cox Utility Services
- The winning jockey in the Kentucky Derby wore 811 – Ryan White and Collin Miyada of USA North 811
- 811 Car and motorcycle built by Paul Junior Designs – One Call Concepts
- 811 wrapped Concrete Truck – NIPSCO
- 811 popcorn bags at high school football games – Enbridge

This is just the tip *of the tip* of the iceberg. These individual, local, high-profile, or annual initiatives all help point people to the same important message. That's essential and must continue. But it doesn't always tell the public or intended audience the full story.

LinkedIn posts with AI-generated images from longtime locate veteran, Jason L. Smith (left), business development specialist at Damage Prevention Solutions, Kelly Carpenter (middle), and lifetime safety advocate Raymond Sonnier (right).

Notable 811 awareness campaigns. Many vendors and service provide help promote 811. Cox Utility Services flew this hot air balloon at events all over the United States (left). Victor Espinoza wore 811 during several Triple Crown races, including winning the Kentucky Derby (right). Photo by Alex Evers.

As we have also pointed out, awareness is only the first step, and more education and ultimately culture is what is needed for permanent safety gains. To do this, an overarching narrative is needed in an effective awareness campaign beyond simply logo familiarity. One campaign that has blended these concepts well has come from an unlikely source. Not only will we have to leave the underground world, but travel to another hemisphere to learn.

Overhead Cable Safety and Protection

One reason damage prevention awareness doesn't always move the needle is that it is a two part message. Telling people to "call before you dig" or even "know what's below" is the first step. But from there, there remains a "why" which is that if you don't, you could hit a buried gas or electric line. That could be deadly. But even though these are small logical steps and are largely common sense, they are nevertheless multiple steps.

Outside the industry, this is done in different ways, Operation Lifesaver, Inc, dedicated to railroad track safety uses "See Tracks? Think Train!" to help communicate the potential for danger anytime someone is on the tracks. But not every industry has such simple direct relationships.

We think about something Roger Cox told us that even with all the slogans, people can witness the process and still miss the substance:

> We have to tell people who we are and what we do ... people see paint on the ground, but they don't know the why.

Awareness that requires linking ideas is challenging, no matter the subject or organization doing the campaign. The public can have marketing fatigue or not feel personally invested in the first or even second element of the campaign.

In Australia, a utility safety campaign elegantly bridged this two-step challenge with a single and highly effective awareness campaign: *Look Up and Live*

Unlike buried power and gas lines, overhead lines are visible. But many people are blind to the infrastructure around them. Very often, truck drivers or heavy equipment operators, who are focused on the task in front of them, do not realize their equipment is actually high up in the air and near a live power line. On average in the United States, two non-electrical workers die every week due to accidental contact with overhead powerlines. Like excavation damage, these are all avoidable.

The passion for safety and preventing damage to utilities is not unique to the United States or to buried facilities. As the conference section above demonstrated, there is keen engagement with damage prevention happening, word play perhaps intended, *Down Under*.

Credit for the "Look Up and Live" campaign is owed to the inimitable Glen "Cookie" Cook with Energy Queensland, the primary electric utility in Queensland, Australia. This campaign is designed to reduce injuries and fatalities from accidental contact with overhead power lines. It emphasizes planning and heightened awareness for workers using tall machinery, trucks, farm equipment, or operating near powerlines. *Importantly, you don't have to make contact with a powerline for it to be fatal.* Electricity can jump if equipment or machinery gets too close to powerlines.

With those stakes, the simple campaign to "look up and live" says it all. Paired with visual logos, this not only raises awareness, but educates about the stakes all in one. From there, it requires a culture of planning and safety before and during the job to prevent incidents.

Cookie points out that: "the biggest reason overhead powerlines are contacted is due to a lack of planning, that's where the lookupandlive.com apps will come into their own. Allowing workers and businesses to adequately plan for overhead powerlines before getting to site." He also

This slogan reminds people to be aware of aerial cables.

makes the point that: "We plan for underground infrastructure we can't see, but most don't plan well around overhead infrastructure we *can* see. That's where inattentional blindness comes in where we don't see obvious dangers, even though they are in plain sight."

Cookie has become the global ambassador for Look Up and Live. He has spoken in the United States, Canada, New Zealand, and across Australia promoting overhead line safety. In fact, he was also chosen as a Damage Prevention Hero of the year for 2022.

Notes on Safety

Working around aerial powerlines requires proactive planning just like working near underground facilities. When planning any job, big or small, always check the location of above ground powerlines to ensure you are able to work safely. If you are working near overhead powerlines, always implement appropriate risk controls.

Look Up and Live's advice is that you should contact the asset owner to see if there is the option to de-energize or relocate the lines or install powerline visual indicators (e.g., rota markers). When this is not possible, always stick to NO GO ZONE rules and implement the following:

1 Placement of physical barriers to prevent encroachment.
2 Appoint a safety observer.
3 Delineate and set up a safety observer area.

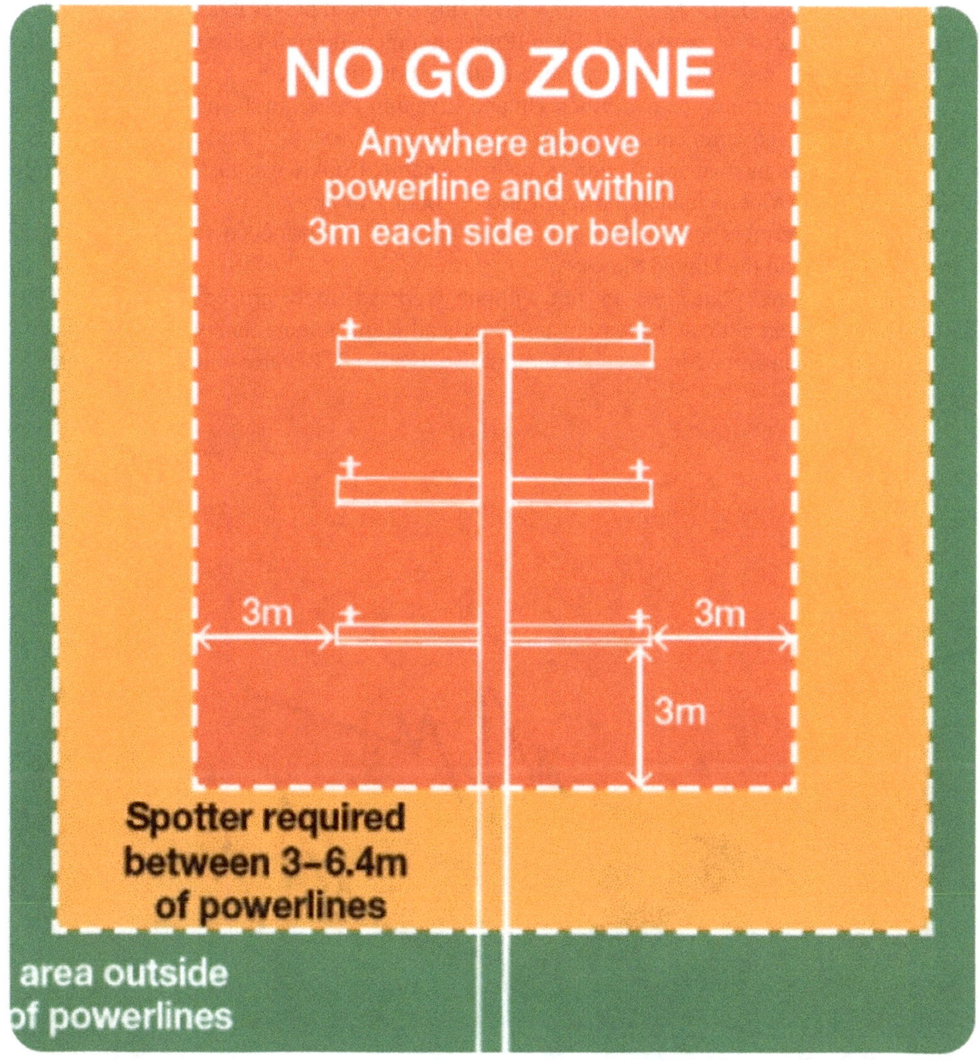

Life saving visual for workers on every job site.

The Look Up and Live app (in Australia) provides an interactive map to identify the powerlines, their voltages, and who owns them. Most of Australia's above-ground electricity network is now active on the map, and the tool is free. Energy Queensland saw the power of this program and eventually transitioned the ownership to Before You Dig Australia, which now promotes it aggressively.

Utility Safety Partners in Alberta, Canada also has an aggressive overhead power line safety program that it acquired when Alberta One-Call Corporation, the Alberta Common Ground Alliance, and the Where's the Line campaign merged under the unified Utility Safety Partners umbrella.

USP education and awareness material stresses to "Learn the 7-Metre Rule": any activity within 7 meters of an overhead power line puts you, and others working nearby, at risk of serious injury or even death. Most incidents involving overhead power lines occur while equipment is being operated or transported. Even if the equipment doesn't make direct contact, electricity can arc or "jump" from the power line to any conductive object.

Alberta is also launching its own Look Up and Live application in the near future and SaskPower, the Crown (government-run) Electrical transmission and distribution company in Saskatchewan, also has its own application by the same name. Cookie was instrumental in helping Saskatchewan and Alberta launch the application in both provinces.

Utility Safety Partners' brand promoting Look Up and Live has been submitted for Trademark for Canada and the United States.

Back in the United States, we are not without overhead safety procedures. In fact, Georgia 811 has overhead notices, by way of the Overhead High Voltage Safety Act established in 1992, which are transmitted like a locate request through the 811 center to its facility members.

The Canadian adaptation of the successful Australian campaign.

The Georgia Public Service Commission requires, by law, that anyone doing work in the vicinity of overhead high-voltage lines must notify Georgia 811 at least 72 hours in advance, excluding weekends and holidays, to enable the owner or operator of the high-voltage lines to determine the precise tract or parcel. Each person who creates a locate request, whether online or via an agent, is asked if the work will come within 10 feet of overhead high voltage power lines. Members have provided Georgia 811 separate map information designating the location of these lines and these tickets are often transmitted to a different department versus those for excavation activity.

As other states engage in overhead safety initiatives or seek to refine their awareness for buried infrastructure, lessons can be drawn from the Look Up and Live slogan and program. While there is strength in a unified and single message that everyone agrees upon, sometimes a new or complementary message is also needed that strikes a different chord or reaches a different audience.

There is no secret recipe for how many or what messages are the right ones. "Call Before You Dig" is a straightforward call to action, which emphasizes the requirement even if the stakes aren't clearly communicated in the same slogan like "Look Up and Live." As awareness and educational efforts continue into the future, it is important that new voices continue to join the chorus. In time, ineffective messages should fade and alignment should form around the best, but there is no harm in trying new things or outsiders carrying the message on.

Third-Party Organizations

In a perfect world, an industry seeking to make every person aware of their work would not only want to leverage fellow industry platforms but entirely third-party organizations and initiatives. If every grocery in America ran an $8.11 deal for all 20-pound bags of charcoal for National 811 Day, millions of new people would be reached during a peak summer barbecue and construction season overlap. This example likely isn't in the cards, but organizations outside the industry (read: not in construction, utility operation, locating, government, or safety processes) can nevertheless be effective partners. Marco's Pizza, which has over 1,200 stores spread across 35 states, has run $8.11 specials during NSDM specifically intended to promote 811. Others from local to franchise to national businesses have also engaged, sometimes once and sometimes with recurring campaigns.

One of the most significant national campaigns is Lowe's Home Improvement branding all trees and plants with the 811 message as well as the branding on their Kobalt long-handled tools. Starting in 2015, this initiative reaches homeowners at the point they are deciding to purchase. It has the benefit of virtually ubiquitous reach all across the nation, getting to consumers in person as they are in store, and even has a validating effect because of the store's brand strength and trust built up over years. Over time, Lowe's, Home Depot, Ace, and other national hardware stores have joined in promotion of 811 messaging. This trusted partner dynamic lends more outreach power and credibility to the message not unlike ambassadors. When name brands can join in, the entire industry is better for it, but that doesn't mean lesser known organizations don't add great value as well.

Groups like Aii, which are truly outside the industry, still work to research and educate the industry itself, public policymakers, and the public about how 811 works. Founded by the first acting administrator of PHMSA and governed by former PHMSA, NTSB, and other high-level policy experts, Aii has spent over a decade at the forefront of infrastructure policy research and education. With damage prevention being central to planning, building, maintaining, and using our nation's infrastructure, it has become a central focus area. Accordingly, over a quarter of the Aii research library is dedicated to research reports and policy briefs discussing challenges and solutions in damage prevention. By joining along with NSDM and 8/11 Day and creating Modern Roots March, Aii helps amplify the message every day.

Well outside of the policy world, many other third-party platforms exist that can be leveraged. Whether as paid partnerships, sponsored activity, or organic passion-based initiatives, the 811 message can be at home in virtually every setting. That includes when it's shared with little children who are used to listening to a giant yellow bird and his other puppet friends. In 1988, Sesame Street featured a hybrid live and animated music video called "what's below the street" highlighting how outside organizations can take the baton and help popularize the ideas, resources, and importance of damage prevention.

Third-party organizations remind us that damage prevention is not confined to a single industry or audience – it can and should be reinforced anywhere people live, work, and build. When outside voices echo the call to "know what's below," the message gains credibility and reaches ears it might otherwise miss. Whether through research groups, public campaigns, or even beloved children's programming, these external amplifiers prove that the culture of safety is strongest when it transcends its own boundaries. With that in mind, the path forward is clear: only when culture, education, communication, and leadership align can damage prevention truly become second nature.

Culture, Education, Culture, Communication, and Culture

The vast majority of damage to infrastructure is 100 percent preventable – even with 50 million miles of buried infrastructure underfoot – and the tools exist to make this a reality. However, that does not mean it is easy or fast. Damage prevention and excavation safety revolve around making the right choices.

As the example Wylie Davidson pointed out with the tree trimers in his yard, culture is critical. Your company has a culture, either by design or because the void which was filled by employees. Culture is not a statement or a mission, it is the actions modeled by leadership. When a stakeholder's leadership team understands the positive ROI on damage prevention along with the moral obligation, the culture should follow. If they believe this in their core, they will fund programs and be visible cheerleaders for the cause. If they truly understand the long-run value, supporting damage prevention is a no brainer.

In 2026, there is no excuse for not knowing the proper steps to prevent damage and dig safely. The organizations discussed earlier in this chapter have all the educational tools anyone could wish for. On top of those national resources, 811 Centers typically provide even more training and education. The 811 Center has localized education that cover those state rules and often address the special needs of the climate and environment in their service area.

Communicating the safe digging message is done by 811 Centers and some national organizations, but the most important communication is the one-on-one, face-to-face communication. If you are reading this book, you are now in a position to educate your coworkers, friends, and family, and to follow Wylie's lead in talking to the people doing the work. Scott talks about the many times he has stopped to talk to someone doing excavation with no paint or flags nearby. The message is always about their safety, not a lecture about the law. When looking them straight in the eye, the message gets through, even if they do not act on it immediately. If you have adult kids, you have likely seen them do things now based on what you told them when they were young, even though it did not appear to be absorbed at the time. Never stop communicating.

Chapter 8 Resources

We encourage to revisit resources from previous chapters, as they are all oriented toward awareness and education in one way or another! Here are a few more to explore:

The CGA's toolkit library

Call 811 Pirate Video – This is a timeless video the CGA created to educate kids between 8 and 11.

ESA/ACTS Now Town Halls

Aii Modern Roots March

Before You Dig New Zealand: Don't Play Around with Your Safety (*This is Benjamin's favorite awareness video)

PHSMA 811 Video Contest – Winning Entry 2017

Damage Prevention Focused organizations and campaigns

- Pipeline Association for Public Awareness
- Pipeline Safety Trust
- North American Telecommunications Damage Prevention Council
- Drain Tile Safety Coalition
- Before You Dig Australia: Look up and live

Podcasts and conferences

- Global Damage Prevention Summit
- ACTS State Damage Preventions Summits in TX, NM, MS, LA, AL, TN
- CGA Conference & Expo
- The Safety Moment by Utility Safety Partners – Stories and strategies
- Coffee with Jim and James – Casual unscripted industry conversations primarily at trade shows and conferences.
- We Are Damage Prevention – a series of short (2–3 minutes) interviews on various damage prevention topics

There are many more conferences and events, we cannot list them all. We recommend that if you want to learn more, start by Googling "811 conference [your state]" and see what may come up!

Notes

1 National Transportation Safety Board (NTSB). (1997, December 1). Protecting excavation public safety through damage prevention (NTSB Safety Study No. SS-97-01; NTIS No. PB97-917003). https://www.ntsb.gov/safety/safety-studies/Pages/SS9701.aspx

2 Pipeline and Hazardous Materials Safety Administration (PHMSA). (2016). PHMSA Focus newsletters, 2002–2010. GovernmentAttic. https://www.governmentattic.org/20docs/PHMSA-FOCUSnewsletters_2002-2010.pdf

3 Safe Excavator. (2019, April 9). 811 TV spot, "micro-Mike Rowe." iSpot. https://www.ispot.tv/ad/IiAK/811-small-mike-rowe

4 Iowa One-Call. (2020, December). Iowa One-Call Excavator Quarterly – Vol. 28 No. 4. https://iowaonecall.com/wp-content/uploads/2020/12/IOC-NewsletterVol28.4.pdf

5 Southern Company. (2022, August 3). Southern Company Gas and National Excavator Initiative get dirty with Mike Rowe to spread the word to contact 811 before digging. https://southerncompany.mediaroom.com/2022-08-03-Southern-Company-Gas-and-National-Excavator-Initiative-get-dirty-with-Mike-Rowe-to-spread-the-word-to-contact-811-before-digging?

6 Southern Company Gas. (2023, June 15). Southern Company Gas Wins Top Industry Honors for National 811 Day Safe Digging Communications Campaign. PR Newswire. https://www.prnewswire.com/news-releases/southern-company-gas-wins-top-industry-honors-for-national-811-day-safe-digging-communications-campaign-301852504.html

Chapter 9

Hurdles and Challenges

Tim did not like to waste time. From the moment he woke up until his eyes were finally shut at night, he was multitasking. Making breakfast? *Morning news playing in the background.* Walking his dog? *Podcast in his earbuds.* Going to the bathroom? *YouTube DIY videos.*

He simply couldn't do anything without doubling up. His time was too precious, and he felt he always needed to get more done than there were hours in the day to achieve. At least this constant stimulus gave him continual awareness and education during mundane tasks so there was no wasted quiet.

His sense of urgency often translated into projects he needed to do around the house – sometimes leading to cutting corners or being distracted. More than once, he had his phone playing audio while he tried to measure something and scratched down the wrong measurement on his notepad.

This weekend, he had some yardwork to do at his grandmother's house. He planned to rush over there in the morning and be wrapped up by lunch time. He knew it involved a few landscaping things – just ripping old shrubs out of a flower bed – and resetting her mailbox that was leaning over and threatening to collapse.

For Tim, this was nothing. He had watched enough DIY videos to know he could do this project in a few hours without breaking much more than a light sweat. And he planned to do it *this* weekend to check it off his list. The only problem was that it was already Thursday and he had a lot to finish at work and at his own house. He didn't want to waste any more time checking in with his grandmother or her neighborhood association or any city officials. It was a simple task on her own property – why bother with red tape and approvals. While he knew about 811, he thought that he wouldn't have to provide notice because he was just pulling up and reslotting a new mailbox in the same hole – no digging, no call!

What he should have accounted for was that nothing ever goes to plan! How many times had he billed out an hour in his mind only for a task to take two. Or took a trip to Home Depot only to return three more times in the same weekend – keeping his head low each new time so the greeters didn't know he was embarrassingly returning *yet again.*

As soon as he started pulling shrubs, a black cable running under the mulch came too. He wasn't sure what it was or what it did, much less whether it was a public or private line. Nothing disconnected or severed, so he innocently pushed it back several inches and recovered it in mulch away from the area he was de-shrubbing.

With a podcast playing in his ears, he moved on to the mailbox, which slid out of its hole with ease. That was no surprise, given the hole was so loose with the leaning post. The new one was ready to slide straight in, which went like silk. He was excited that he would have this wrapped up in no time. He packed some extra dirt into the hole and an expanding foam agent that secured the post – which

DOI: 10.1201/9781003796619-11

he'd seen in DIY videos but never used before. It probably wasn't as good as quikrete, but a little old lady's mailbox didn't need the kind of foundation that a fence post or deck structure needed.

The last thing he had to do was "pull some weeds." They were along a side portion of the yard and his grandmother had assured him it would be simple. She even had a nice young man come and spray them last week, so they were already desiccated and brown. Pulling them up would just take a few minutes. The sand-colored weeds included a thick coverage of St. Augustine grass as well – now similarly dry and discolored. It was in a patch of yard that was supposed to be landscaping rock, but all the river rock pebbles and smooth stones had long since been consumed by the lawn and weeds.

As Tim got to work pulling, he realized that dead weeds did not pull up as easily as he hoped. Long, dense stems and stolons ran dozens of inches in all directions just beneath the surface of the blades. He couldn't pull up large patches at one time because the roots broke off. Frustrated, he went to the shed to pull out a new weapon – a rake. This, he thought, would solve two problems at once: hook into the weeds and grass to uproot them and help bring the landscaping stone back to the top and distribute it well.

Within three heaves, this plan was scrapped. He went back to the shed – so focused he was paying no attention to the droning podcast voices and instead was engaged in a fierce dialogue of his own inside his mind about what weapon to wield next.

He returned with a flat edged shovel and inserted it along the line of grass. This worked like a charm, both severing the stolons and gliding underneath the shallow root layer. As he went, he slipped deeper and deeper without realizing it, riding on the momentum of his recent success and hoping to get this project over with. He soon found himself outright digging when he only ever meant to be pulling weeds or at most raking or surface scraping.

It wasn't long before his shovel met interference. He felt tension as he maneuvered it and assumed it was a nasty root. A few minutes later, he noticed it was stark orange. But his weed work was done, so he piled all the uprooted stems and leaves into a bucket to take back behind the shed.

A few minutes past that, a frustrated neighbor came out wondering why he had lost internet connection. Tim was out back returning the shovel and pitching the weeds behind a fence, and the neighbor went back inside. When Tim came back, he contentedly surveyed his work – the now de-shrubbed flower bed, a newly installed mailbox, and landscaping stone that was visible again, albeit caked in dirt as they were clearly resurrected from the subsurface and not freshly laid.

Tim was not alone this morning. Across the street, he could hear three kids playing at a small park, shouting and laughing with their games. They had been out for a while and waved at dozens of cars and dog walkers as they passed by.

The oldest was 11 and she was the fearless leader. Her younger brother was about 7, and he did as she instructed. The third child was a boy about 10, who had more energy than both siblings put together, but still followed the girl's lead.

After a third game of tag in a row with little brother being *it* for over five minutes, his sister proposed a new game – *flag tag*! The neighbor boy's eyes lit up, he knew just where this was going. He sprinted across the park and yanked up three bright yellow flags he had seen all in a row. He proudly distributed them to his compatriots, and they went about parading with the little flags, trying to tag each other and poking each other with the thin metal rods. When the game's charm wore off, they simply discarded them haphazardly on the ground in the park. A well-meaning nanny who was watching a five-year old in the park saw them and thought it was litter, so she relocated them to a trash can.

The three children walked back to their homes across the street as the sky grew cloudy. Before making it indoors, the ring-leading girl ran past and slapped a yard sign at a house along the way. Her little brother dutifully did the same. It was a for sale sign for a nice little cottage house. The neighborhood was lovely, with the little park and several families. The sellers were hoping they could get the most out of the sale.

The only problem was that the sidewalk in front of their house looked like it was graffitied. The wife was distraught the morning she saw men spraying yellow, red, and blue on the street, sidewalk, and a portion of her own driveway. What's worse, it extended well down the street. Who would buy this house she had worked so hard to prepare to put on the market? All those hours hand scrubbing baseboards, matching the wall color for touchup paint, and making sure the windows were streak-free with windex and the screens were cleaned off with a vinegar solution – all for naught in her mind as her curb appeal was destroyed with lines, dots, logos, and signatures in spray paint just inches past her manicured lawn.

The further down the street the paint went, the clearer it became that it is not a simple homeowner project. Something much larger in scale was planned. In fact, what the home-selling wife did not know is that locators had been to the site three times already, and yet no construction had begun in the neighborhood. Two streets away, an excavation crew was doing some kind of work she couldn't identify. Maybe it was updating the city water infrastructure, maybe building a new sidewalk. Whatever it was, the request for locating and marking was extensive and much larger than the crew was ready to conduct work on. But not only that, it was buffered so large that it included side streets and homes like hers even though the eventual work would not really be near them.

Locates and Tickets and Delays ... Oh My!

The damage prevention industry has been standardized over the course of decades. The process of notifying 811 at least two days ahead of excavation, locating the site, and safe excavation respecting the marks is as clear and formalized as any other standard operating procedure in any other industry. Yet damages persist, tensions between stakeholders grow, and inefficiencies within the process grind like nails on a chalk board. Stated simply, the process works and it doesn't.

Perhaps no single quote can better encapsulate the state of the industry than Charles Dickens's seminal introduction to A Tale of Two Cities.

> It was the best of times, it was the worst of times, it was the age of wisdom, it was the age of foolishness, it was the epoch of belief, it was the epoch of incredulity, it was the season of Light, it was the season of Darkness, it was the spring of hope, it was the winter of despair, we had everything before us, we had nothing before us, we were all going direct to Heaven, we were all going direct the other way – in short, the period was so far like the present period that some of its noisiest authorities insisted on its being received, for good or for evil, in the superlative degree of comparison only.

It depends on who you ask, or on what day you ask them, but the damage prevention industry has reason for incredible optimism and room for substantial improvement. The key point illustrated in this chapter, however, is that while disruptive reforms can and perhaps should occur in some areas, significant improvement can be made with simple, modest refinements that do not substantially alter the current process or require starting the industry back at square one.

To tell this story, we enlisted help. It did not seem right for Scott and Benjamin, despite our unique vantage points and expertise, to tell the industry what its problems are or how to solve them. That is inside baseball, and we respect these men and women too much to come in and tell them how to do their jobs. So instead, they told us. And now they are telling you, dear reader.

Through interviews and surveys and even live stakeholder participation during the 2025 Global Damage Prevention Summit in Dallas, Texas, we collected the following feedback. We are placing it here in the book because by this time, you will have read all about the logistics and history of how the industry works and came to be. You also read what recent and ongoing initiatives are aimed at building a safety culture within while proclaiming awareness and education outwardly.

Now, you'll read some of the key challenges just before turning to the chapters that highlight how the industry is already working on solutions and what other options exist to enact reforms.

While the survey does not comprise 100 percent of the sources of feedback for this chapter, we thought it was important for transparency to show you which stakeholder voices were most represented in that process. Many dozens of interviews bolster and support the survey responses, as well as expert peer reviewers for this chapter and full book. But it is also important to note that this chapter breaks the mold from how the industry usually progresses in some respects.

Particularly within the Common Ground Alliance, progress moves by consensus. That means stakeholder groups work together, propose ideas, iterate, and agree. This chapter is the wild west

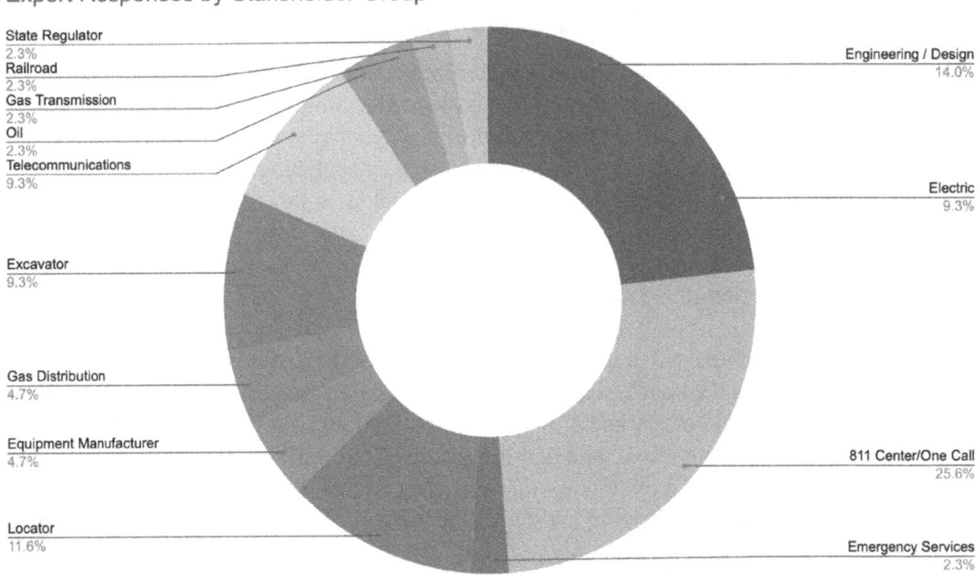

Survey feedback received responses representing 13 of the 16 stakeholder groups, collected September 2025.

of feedback. But rather than young gunslingers shooting from the hip, it is veteran sheriffs firing with pinpoint accuracy. They are speaking for themselves and saying what maybe they can't always say in committee meetings, or espousing ideas they know won't be consensus, even if they benefit the whole industry. To advance ideas most candidly, we've elected to showcase quotes without attribution. So, let's dive in.

The top five themes that emerge, in order, are:

1 Locating, Mapping, and Data Accuracy
2 Communication and Coordination Gaps
3 Data Integration and Technology Adoption
4 Regulatory, Structural, and Incentive Barriers
5 Cultural, Leadership, and Workforce Issues

Commonalities from the raw data also indicate:

- *Interconnectedness*: Nearly every comment touches multiple layers – technical, regulatory, and behavioral – showing that improvement requires a systems approach.
- *Familiarity with limitations of the Status Quo*: Phrases like "no feedback loop," "reactive ecosystem," and "rigidity of system" convey areas of stagnation despite widespread awareness.
- *Readiness for Reform*: Many respondents propose forward-looking solutions – integration, certification, single source of truth, technology alignment – indicating readiness for change when leadership and incentives align.

Locating, Mapping, and Data Accuracy

The challenge of mapping and creating accurate records is a prominent challenge voiced across the industry. It has come out with light criticism, overwhelming concern, sharp critique, and even optimistic opportunity for reform. One humorous way a contractor explained it is that SUE should ultimately help update and lock in place a detailed knowledge of what's below, and when new lines are added, they should be built into a high quality database:

> These utilities didn't just appear in the ground, we put them there and forgot where we put them. It's like a chef who went grocery shopping, loaded the fridge, and now calls in a guy off the street to remind him what's in there.

A frustrated contractor may or may not represent the entire industry, but it makes an important point: the way we've allowed new infrastructure to be built and forgotten about cannot continue, and it is the responsibility of the damage prevention industry to ensure they have (or *can* have) the accurate and precise location of everything below ground at a given site in real time.

In fact, this very idea has bounced around the industry in many ways. Another stakeholder shared that,

> I hear a lot, and it just might be me, that a 'One and done' approach can be made in regard to digital mapping – a locator goes to a site, uses technology to immediately upload to the Cloud the 'exact' location, within centimeters, of the utility, possibly negating the need for future locates if map sharing is allowed.

Where these utilities exist and where locators are sent matters. Because no two locate jobs are ever the same, it takes skilled technicians to know how to handle each environment. But some experts caution that a breakdown in the entire process can occur in a very common but overlooked environment: water.

> The top hurdle in the industry is getting facility owners and operators to understand, they have more assets in waterways than they think they do.

This matters because state laws often do not even consider marine contexts and force the standard 811 process for land-based utilities to apply for submerged pipelines, fiber optic cables, and more. That's a big issue. A land-based tolerance zone may work for a distribution pipeline under a sidewalk, but it does not apply to marine dredging equipment. Likewise, excavators may not think they need to contact 811 at all for water-based activity, locators are unable to spray paint on the water's surface, and many activities outside of traditional excavation (e.g., pile driving or anchor dragging) can pose serious risks to pipelines and other lines.

Regardless of where the facilities are, having them mapped digitally and communicating that to locators is essential. Stakeholder feedback demonstrates this with responses like

> "There is no feedback loop." and "The current damage prevention ecosystem is reactive. We make an educated guess what is in the area before we work rather than document with accuracy after the work."

> Another major challenge is the industry's slow adoption of technology. We already have tools to digitally capture facility locations, but time and cost concerns keep this from happening, leaving us behind in advancing damage prevention with more accurate maps and future innovations like augmented reality.

These flow naturally into the work and challenges locators themselves face. While earlier chapters already discussed workload, scheduling, and pay challenges, there are other factors the experts point to that persist.

The execution of a good locate is dependent on many factors. While the most common issues are work load and training, Jim Anspach point out some of the things he has seen during his many decades in the industry:

> I have always believed the 4 largest challenges are traffic control (or lack thereof), inability to place the transmitter or leads where they need to be (as in a manhole or busy street), unwillingness to use induction, and uneven workload, mostly due to weather disruptions.

One 30-year veteran of the locating industry put it this way:

> The biggest challenge in utility locating is the lack of a structured schedule that ensures locate requests are given enough time to be completed. To address this, there must be a clearly defined limit on how much one locator can reasonably complete in a single day.

Along with "low pay," "uneven workflow," and "lack of consistency and best practices" for locators across different state contexts, there is clear room for improvement. As the industry strives for more accurate maps and faster locates, the challenge expands beyond tools and training to the people who must communicate and coordinate every step of the process. Accuracy on paper means little without alignment in practice.

Communication and Coordination Gaps

Because the entire process and all parties depend on good maps, it makes sense that mapping and locating challenges are the top hurdle on the minds of the experts. But just as core to the mission of damage prevention is communication and coordination between parties.

> One of the biggest challenges the industry faces is the persistent lack of direct communication between excavators, locators, and utility operators. As an 811 center, we can facilitate the exchange of ticket information, but the real collaboration and clarification often needs to happen outside the system. Despite years of recognizing communication as essential, gaps still remain, and those gaps continue to drive risk in the field.

Communication is the most effective when all the stakeholders who are directly involved in an excavation project start out with a solid understanding of how each stakeholder's part impacts the others. When a contractor knows that locators are at maximum capacity, they are less likely to request locates for more work than they can do in the required time. When facility owners, locators, and excavators all know the challenges and understand each other's role, they are more likely to have face-to-face meetings, use white-lining, and provide extremely accurate information about the project.

This is underscored by the desire for real "Trust on all sides of the equation." Without trust and transparency, all stakeholders may not be able to have confidence in the system and may not buy in as fully as they can. If every party withholds their full commitment, gaps emerge, and incidents happen in the gaps. Better communication can close many gaps, but technology can make those conversations more efficient and data-driven.

Data Integration and Technology Adoption

Some gaps may be due to reliance on older processes, even when new tools and technology are available. In fact, a challenge one respondent elucidated implies the very availability of the right tools: "Lack of adoption of modern processes and tools." The fact that they exist but are not being utilized systemically leaves things on the table: cost savings, safety, and efficiency. Experts indicate these can be addressed but it takes deploying new tools, embracing technology, and leveraging data. Phrased as responses to the biggest challenge currently facing the industry, responses included:

> Fast-tracking and integrating innovative disruptive beneficial technologies within the tried and true established approaches to meeting future demands in a secure and resilient manner.

> Identifying viable ways to integrate emerging technologies (XR, AR, VR, AI) in damage prevention – without blindly embracing technology buzz words, but rather better understanding the possibilities, and limitations, of technology.

> Combining and integrating multiple technical, industrial, and government stakeholders in a 'Manhattan Project'-like focused approach to make this happen, rather than the slow awareness and adoption traditional approach.
>
> Proper use of data in the industry. Much is incomplete, lacks validation, and buffers bias and assumptions.

These and other responses demonstrate not only an appetite for utilizing technology, but a recognition that failing to use it is costly. The desire to integrate more technology sheds light on a remaining hurdle being an industry-wide desire to retain the comfortable status quote, even if it is relatively analogue. Nevertheless, even the best tools are limited by the systems and rules they operate within.

Regulatory, Structural, and Incentive Barriers

While mapping and locating were the clear top challenges, industry veterans had the most to say about the system itself. Overwhelmingly, responses touched on laws, regulations, standard operating procedures, and more as representing ongoing hurdles in need of constant overcoming.

At the simplest level, stakeholders pointed out that "laws are not uniform across country." while others pointed to the very laws themselves being too inflexible: "Rigidity of system (laws, regulations, etc.)."

The insights about how the system is set up centered on how those rules reward or encourage certain behaviors. This was regularly cast around "incentives."

> Incentives are higher for shortcuts over safety and damage prevention.
>
> (financial) incentive alignment.
>
> Misalignment of interests.
>
> Misaligned and missing value incentives.

To address some of these, one veteran proposed that:

> For legislation to truly serve damage prevention, it must include input from seasoned boots-on-the-ground veterans that have no vested interest in the cost of utility locates. Without their perspective, the focus shifts away from safety and accountability and instead becomes about protecting stakeholder profits.

While laws can set standards, only people can sustain them – through leadership, accountability, and daily commitment.

Cultural, Leadership, and Workforce Issues

Finally, as we emphasized in the previous chapter, not all solutions require new laws. In fact, the most effective type of change is grass roots and cultural change. While that should be a top-down standard, with leaders defining and cultivating a dedication to safety, it must be found at every job site and recognizable at the ground level. Two responses help demonstrate how these challenges look to stakeholders and what clear roadmap they imply about the solution:

> Getting the damage prevention community to understand they need to protect 100% of their assets, not 95%.

> Education and learning to be proactive.

Surprisingly, most survey responses, while diverging in the particulars, were highly aligned on the broad strokes. They centered around similar challenges in similar rank order, regardless of whether they came from locators, 811 centers, or utility operators. That is an important finding in its own right and implies that most parties see the same challenges and hurdles, just from slightly different angles. That may mean aligning reform efforts can happen with the right language and opportunities. A final observation ties together many of the threads at once:

> Branding, slogans and swag – from coasters to monster trucks – have their place among delivering effective communications but are meaningless and powerless unless anchored to effective procedural change, technology, and method. Setting damage reduction goals won't happen unless there's a roadmap to achieve them. In my view, the damage prevention industry in many places has lost its way by willfully ignoring real change and improvements around the globe and even preferring to focus on bravado and style over substance. Significant change is needed among damage prevention leaders, boards, and governance to focus on improving technology, data, and bridging the gaps that will facilitate ease of public use.

Importantly, this and other comments leave room for just where the reforms can come from. They can be voluntary initiatives, company-level undertakings, or even state or nationwide legal reforms. In the United States, the legal set up is itself both a hurdle and opportunity. The federal system means things can be refined to work best given a state's specific context and preferences, while also enabling states to learn from one another and see how legislative and regulatory reforms work out over time. All this diversity can be a challenge, but it also affords ample opportunity for learning.

50 State Laws. 16 Stakeholder Groups. No Problem!

In some ways, damage prevention is like politics in that almost everyone wants the same outcome, but the devil is in the way people want to get there. This industry is full of professionals on a mission to reduce damage and make excavation safe, but some may still lose sight of priorities on the business side of the equation. Obviously, safety trumps everything, but that does not mean the bottom line should not be considered. Safety and efficiency can go hand in hand, especially when technology and people work together.

The industry has done an amazing job coming up with solutions to problems over the years. In this chapter, we merely outline some of the challenges the industry still faces. This is not a list of complaints, just real world obstacles the industry needs to get past. Fresh ideas, new technology, continually improving communication, with a focus on saving lives will keep the ball moving forward.

As the previous chapter made clear, the industry has many solutions in play. But even while working on all those solutions, there is ample opportunity for new ideas, refinements and reforms, and even systemic changes.

Note to Self

After naming their frustrations and hopes, many respondents offered advice to the next generation – personal notes distilled from decades of hard-won wisdom in the field. Their reflections, presented below, form both a mirror and a map for where the industry can go next.

- *This is the best industry to work in. Everyone is here for safety and damage prevention and though we may have differing opinions on some things, we're all working towards the same objectives. Make the biggest impact you can on improvements to the system.*
- *Take time to understand the perspectives of all the people involved; owners, contractors, utilities, and the designers/engineers. Don't get caught up thinking the technical tools are the hardest part; the real challenge is aligning incentives and trust between parties. Pay attention to how value is defined and measured, because that's what shapes behavior. And remember: progress often comes slower than you expect, but it builds if you keep nudging the system in the right direction.*
- *The key to safety and damage prevention in excavation is relationships with empathy from all stakeholders. To be successful, think beyond yourself. Find the key stakeholders and build a strong relationship with them. Spend time getting to know safety and damage prevention from their point of view. Be creative and help find a middle ground that will help all involved to achieve their goals with the highest level of safety and damage prevention.*
- *Always challenge the status quo. Push through barriers. Find the big picture and lead people to it.*
- *Be driven, creative, and analytical in your desires to provide solutions that help humanity. But don't stop there. Identify all the key stakeholders that will be needed to open doors, drive implementation, and creatively troubleshoot to remove hurdles to adoption and successful implementation. Don't wait for others to act. Reach out and become a loud voice to all who will listen. Advocate, embrace, and help expand the team where everyone can win and see value, moving forward together.*
- *This industry is more political than it might first appear. It is filled with truly passionate individuals who strive to prevent damages – saving lives, protecting property, and insuring the un-interruption of essential services. At the same time, there are political agendas in place that are not necessarily aligned with damage prevention, but more focused on growing power and possibly finding increased business. Advice? Be mindful of the politics. Watch closely, and listen even more closely.*
- *Follow the procedures, they are there to protect you, the crew and facilities. Trust the information you are given but always take it with a grain of salt. When in doubt, dig by hand and take your time.*
- *Slow down and respect the process, because rushing is where mistakes, and damages, live. The marks you put on the ground aren't just paint, they're the difference between chaos and safety. Every storm you walk through will shape your grit, and every mistake you own will sharpen your skill. Trust your instincts, invest in learning every day, and remember, this job isn't about checking boxes, it's about protecting lives. Damage prevention isn't just a career, it's your purpose.*
- *Stay patient and keep learning, because every lesson – whether from success or mistakes – will shape you into a stronger professional. Don't be afraid to ask questions or lean on others; relationships, trust, kindness, and collaboration are just as important as technical skill. Remember that the work you're doing isn't just a job – it protects lives, keeps communities running, and makes a difference every single day. Everyone wants to come home to utilities that work; you will reap what you sow. You are your brother's keeper in the end.*

- *Experience in the utility industry and your love for public power will lead you down a path to 'Damage Prevention.' Being a damage prevention professional will provide you with the fulfilling career you desire. There will always be change and challenges and the hurdles of the industry will keep you motivated to help and stay engaged.*
- *Don't underestimate the value of data and technology in this industry. Stay curious, learn the tools sooner, and don't be afraid to speak up about how data can change the way we think and work. Encourage others to be open-minded and forward-looking, because solutions will come faster if we all start asking the right questions earlier. The sooner you embrace this mindset, the more impact you'll have down the road.*
- *Educate yourself on AI and accurate reliable mapping, since AI will be replacing field work except for very special situations. If you are highly technical, learn about advanced geophysics. Most of all, learn everything you can about utility company operations.*

Together, these voices remind us that the industry's greatest strength lies in its people – their integrity, ingenuity, and resolve to keep improving. What comes next builds on that foundation, translating wisdom into innovation and progress.

Chapter 9 Resources

ESA/ACTS Now Town Halls

Global 811 Magazine

Planet Underground TV – A YouTube channel packed with damage prevention and locating videos including Locator Training, Digging Dangers, Roundtables, Industry Shorts, and more.

Chapter 10

Breaking New Ground

Data, Technology, and Talent

With his fingers moving faster than his mother's eyes could track, Kevin twisted and spun the rubix cube like it was second nature. He had played with them since he was a kid, and now at the strapping age of 18, he was nothing short of a master. Problems and puzzles simply unraveled themselves for Kevin. He could see things differently than any of his siblings, and certainly than his parents.

In school, he was a straight B student – not for lack of talent, but he got bored easily. If he wasn't being challenged, he may drift off in thought or engineer something out of his backpack straps, pencils, and zippers while the teacher droned on from the front of the classroom. In fact, he aced every test, but often forgot to bring in homework or complete small tasks.

When he was focused on a problem, though, Kevin was unstoppable. The question in front of him – and his parents and school guidance counselor – was what he should do next. College? Silicon Valley? Entrepreneurship?

Kevin taught himself code in his spare time – or read books about it in class when he should have been listening to a lecture on cholera outbreaks or Shakespeare. He built his own computer at home and optimized the settings and setups for his parents' computers, too. He really loved the interplay between a physical challenge and a virtual opportunity – and that's where computers really came to life for him. He could design hardware and then write software for an end-to-end solution to a problem.

When a college and career fair came to his school, he did a scan across the room to decide what was worth walking by. The military recruiters were a hard pass, as was the local community college, an animal shelter, and the municipal waste service. As he slipped through his peers and ensured he avoided eye contact with the uniformed Army representatives, he passed something that caught his eye:

"Help Us Save Lives Through Data"

It wasn't exactly his inner altruism that stopped him, but the relationship between lifesaving and data that piqued his curiosity. He lingered long enough for the table sponsor to ask him his name. After quietly replying "Kevin," he bluntly asked, "So, what do you do?"

The chipper young woman was so glad to be asked this question. She likely rehearsed the answer to herself and delivered it hundreds of times by now, but each time it came off with the same authentic energy as though it was her first.

"We receive calls and online requests from people doing digging work and use our mapping software and computer databases to notify utility companies before their pipelines and fiber get damaged," she said confidently, working just enough buzz words in to ensure a follow up

DOI: 10.1201/9781003796619-12

question. And he did have follow ups. In fact, Kevin stood there for a full five minutes – an eternity in high school reckoning – asking about datasets, virtual maps, artificial intelligence, and more. And after five minutes, they still only scratched the surface of the innovative tools already in use or the new systems and processes yet to be created that will come into being in the next 10 years of the industry.

High Stakes, High Tech

An industry that involves moving dirt may not sound like the most high-tech sector. But the damage prevention industry is an interconnected ecosystem, not only of different stakeholder groups but also layers of technology. Even the basic one-call concept of calling 811 or using an online ticket involves telecommunications networks and servers, these rely on fiber optic infrastructure (the very same fiber optic infrastructure at risk of being damaged, no less), and more. Once the request has been made, the facility notification center relies on a multilayered digital map to find overlap between the proposed excavation site and member utility company's lines. Some of these also utilize satellite imagery, tying this underground industry directly to the ultra-high-tech space industry. After consulting the virtual map, locators are sent to the scene to use sophisticated technology to peer into the subsurface.

More and more, these processes are being bolstered by drones, LiDAR, augmented reality, and additional technology. Artificial intelligence is helping improve map quality and incorporating abandoned lines back into the databases. And digital twins are being generated so that professionals at computers can visualize the entire underground miles away from the site with pinpoint accuracy.

Some of these tasks are taking place in real time, but they also relate primarily to the process itself – that is, receiving a request, processing the ticket, and having locators mark the site to be excavation-ready. Once the process has occurred several thousand times, the Facility Notification Centers then have volumes of data they can analyze, which they can then use to refine and improve their operations. These optimization problems will ultimately save billions of dollars across the country, while also critically saving lives.

Locator companies similarly have to optimize their operations for both financial and logistical success. Incoming requests, finite workforce, and varied state laws mean that one single company operating in multiple states will have new and evolving challenges to solve on an almost daily basis. A single schedule or standard operating procedure cannot keep pace. New algorithms, incorporation and training of machine learning, and AI will be essential for continuous improvement.

Getting Ahead of No Call Damage

As we mentioned in the chapter on utilities, there are forward-leaning opportunities to use the very facilities themselves as a detection asset. That means that rather than hoping every stakeholder does their job perfectly, we can leverage the fiber optic cable already in the ground to listen to what is happening in rights-of-way. Because nearly one-third of excavation damages are associated with no notification being given, this is a pivotal opportunity to get ahead. For an industry aimed at being proactive, this is the bleeding edge of proactivity.

Even a single strand within a fiber optic cable can be repurposed as distributed sensors. By sending laser pulses, then deciphering how the light behaves inside the fiber, asset owners and 811 Centers can detect digging, excavation, manhole openings, tapping, saw-cutting, gunshots, leaks, or explosions in real time. This allows 24/7 live monitoring when there is a fiber in place to utilize this technology. Excavation activity can be sensed up to 30 feet away.

The process works hand in hand with sophisticated interrogator equipment and algorithm-based analysis. Volumes of data that are meaningless unless interpreted – indicating everything from noise to vibrations to temperature or pressure changes – can be categorized to precise activity types. That means the fiber can "hear" the difference and tell the organization monitoring the data the specifics between a heavy truck driving past a utility corridor and someone using a jackhammer several feet or meters away.

When this data is incorporated into the Facility Notification Center's operations, they can cross-reference it with open tickets to know if someone has followed the law and all the utilities have been marked. In that case, they can report to each company that the work has begun. If no ticket is found, the 811 Center can coordinate to dispatch a team or notify the utilities about potential threats in real-time and stop excavation work before it creates damage.

> The same fiber that's used for telecom can be turned into a distributed sensor … able to detect if someone is tapping with a shovel or digging with an excavator … and identify within 5, 10, 20 feet what is happening.
>
> (Paul Dickinson, Founder and Principal Smart Infrastructure Solutions, LLC)

This type of distributed fiber optic sensing is a relatively new tool in the damage prevention toolbox, and it can protect all the facilities in the right-of-way. In many cases, there is existing dark fiber (unused) that could be put to use without any new fiber optic infrastructure needing to be laid down. That means millions of miles of utility corridors and hundreds of billions of dollars of infrastructure can all be protected now without new digging, cameras, sensors, or reforms to the system.

> These systems can tell the difference between a jackhammer, an excavator, or a saw cutting concrete. They can accurately locate and classify disturbances along a utility path.
>
> (Michael Montgomery, AP Sensing)

To date, this fiber solution has not seen wide adoption. It is a proven technology, but more stakeholders have to know about it for it to see systemic implementation. From there, it will take sophisticated and visionary leaders with chops in data analysis, coding, computer science, engineering, and more to refine and hone it. This is one of the most promising interplays between hardware and software to make the biggest impact for safety in a generation. Young talent, working with industry veterans, will make it a reality. The next challenge innovative thinkers are already tackling is the updating and improving of records and maps.

Digitizing and Improving Mapping

Technology is moving mapping from outdated records Quality Level D (the lowest level of accuracy under ASCE 38) toward real-time, GPS-accurate, shareable, and 3D-enabled maps. This shift is driven by the increased use of SUE, Real-Time Kinematic satellite-based positioning technique (RTK), and 811 Center data integration. RTK creates very precise GPR, which can be within half an inch. This convergence of technology advancements is expected to cut damages, reduce downtime, and greatly improve excavation safety.

James Wingate, Executive Director of USA North 811, describes how one of their members has reduced damages using this data. This is especially encouraging because water utilities, on average, are not the most aggressive damage prevention stakeholder group. This sets a great example for others.

> Since GPS'd, this water district has located 50,000 tickets with 99.9% accuracy. Any utility operator would love that kind of data.
>
> (James Wingate)

This and related work are only possible because of innovators within and outside the industry. While damage prevention professionals have long created better processes and procedures, the story of rapid improvement in recent decades has been facilitated by outside vendors, technology companies, and people willing to think about old problems from a different vantage point.

Among many other companies working on this problem, 4M Analytics has taken on the challenge of big data and made it manageable. While having immense volumes of data may sound like a good problem, it can also be overwhelming and actually halt progress (as a lawyer, *Benjamin here*, I can attest that a "document dump" is often a type of legal warfare meant to paralyze the other side. This can also be known as malicious compliance, where one fulfills a request in an unhelpful way, like paying a big fine in pennies.) In the old days, it took immense manpower to read documents, analyze and categorize them, and discern learnings from them. Today, this is all done by computers.

4M describes itself as the "first Utility AI Platform that compiles millions of scattered public utility records into one easy-to-use map." They are using AI like a lasso to make huge datasets manageable, tying together relevant details, and forming a cohesive and coherent map of the underground. If you can picture Facility Notification Centers having a decent understanding of what is in most places of a state because the major utilities have to submit their facilities in general geographic terms, then 4M and companies like it, are forming their own digital twin to display virtually almost exactly what really exists underground. They aren't using polygons or buffers, but using real records and existing maps, refining them when new data becomes available, and looking backwards for historical details to hone the data into a clean and crisp representation of what is below.

FuzionView, mentioned as well in the chapter on SUE and Mapping, is likewise pulling data from many sources and compiling it in real-time for usable insights. It is designed to work across different data owners and mapping protocols for maximum effectiveness.

> The engine has been designed to work with and number of disparate data owners of various online mapping services such as WFS and REST to name a couple and can easily be added to via it's modular implementation framework. All FuzionView services are API driven and use open standards where appropriate, so the integration of FuzionView engine into existing systems requires a very low amount of effort on a developer's part.[1]

As the industry and its partners push forward on these projects, not only will they dramatically improve our understanding of the underground, but they'll eliminate the *chef who forgot what he loaded in the fridge* problem. In time, old records will be updated and digitized, and new computer applications will turn those records into more valuable products and platforms.

The need for GIS, computer science, coding, and other disciplines is higher than ever. Whether from Silicon Valley or from a local 811 Center, increasingly young talent can dramatically improve safety and damage prevention from their keyboards.

Crunching Data, Saving Lives

AI and data will absolutely change and improve damage prevention efforts moving forward. Being able to access and use more data can also create complications. It turns out that for years, excavators, locators, and utilities interpreted "on-time completion" of a ticket differently. Utilities and locators believe that entering a positive response code before the deadline signified completion. Excavators felt "completion" was when they were able to legally start work, often based on the day count. While this may feel a little in the weeds, this isn't just inside baseball, it's a major challenge that goes to the heart of how 811 works.

The stakes are significant for everyone involved – from first-time excavators to industry veterans. Even a minor misunderstanding can delay projects, create safety risks, or lead to costly damage. And because each state may define and handle these terms differently, contractors who work across state lines face an even greater challenge. Clarity and consistency in how "completion" is understood are therefore not just administrative details – they are essential to keeping workers safe and the system functioning as intended.

For Facility Notification Centers, this is one of the new major issues to resolve, and they are doing so with data. Spearheaded by Georgia 811, elevated by CGA, and followed closely by Virginia 811, the topic of excavation readiness required statistical analysis and modeling. Taking in volumes of incoming tickets and assessing the dozens of positive response codes, these states looked through years of data to understand how often a site was truly ready by the time state law said it should be. That means after an excavator requested a locate, after a ticket was created and all relevant utilities were notified, after all locators completed their markings, after all positive responses were posted, after any unsettled issue or altered marking schedule was completed, and the legal timeframe was completed.

> Most members and locators were using on-time completion to describe the act of making sure a PRIS response was in by the legal date and time. Excavators were using it to mean we can start work.
>
> (Meghan Rafinski, President and CEO, Georgia 811)

With new software tools, Georgia 811 decided to determine what percent of tickets were ready for the excavator to begin work by the due date on the ticket. They coined the term "excavation readiness" to mean all final responses were in and excavation can begin. On the surface, this seems simple, because positive response is part of the Georgia law. One tricky part of this analysis is the fact that there are 38 different codes for positive response. A number of the codes indicate that everyone is doing their part, per the law, but there may still be a reason that digging cannot begin.

Georgia 811's extensive evaluation of their data resulted in a formula factoring in ticket type, cancellations, disputes, and disqualifying codes in order to come up with a way to measure the excavation readiness. It has become a blueprint for other states.

> If one of us figures out something valuable and then we share it with each other … we've just got to keep going. This is absolutely just the beginning.
>
> (Meghan Rafinski, Georgia 811)

Any miscommunication in this industry can literally have life and death consequences. That means realizing stakeholders perceived a term differently and ultimately resolving it is paramount. The fact that it is also a statistical data analysis problem just shows how complex and specialized this industry is as well. Virginia 811 was next to complete their own analysis. With each state having its own number and type of positive response codes, it is unlikely there will ever be a national standard, but the more states that complete this analysis, the more each can learn from the others.

> Every state's going to have to look at this differently … There can be a notion we need to compare state to state, apples-to-apples. That just really isn't the reality.
>
> (Kenny Spade, Manager of Data Analytics & QA/QC at VA811)

Once all this analysis is complete, it can inform state-level reforms or refinements. It likely will not mean one state is superior to another, but each has its own legal and cultural specificities, as well as geographic factors, climates, and more. This richness makes the problem more interesting, because every state will have to have their own specialized statistical framework, their own data analysis, and their own solutions to new and emerging challenges. That signals lots of work ahead, both for the industry to continue chipping away at and for new talent to come in and tackle!

Taking Data, Telling Stories

Finally, even with this data in hand, there is still more to do. Collecting and analyzing information is a skill in itself, and when it is passed along, it requires a different skillset to make value from it.

> Data is not fact. It is something that is up for interpretation. The language we use around data … can drive decisions. We need to be careful and step back and make sure the decisions we're making are truly aligned with a well-vetted database.
>
> (Scott Crawford, President and CEO, Virginia 811)

Not everything the damage prevention industry needs is in the ultra-technical, AI, or coding realm. All of those things are rising in the industry, but there is also great need for people to turn the learnings, improved records, and technology into a compelling narrative. Not only is this needed to continuously improve awareness efforts for the public, but it is essential for stakeholder education and even for policymaker outreach. Without clear language, good communication skills, and a story to tell, much of the essential work being done will not make it to the right people.

Better marketing, awareness, and education are evergreen needs in the industry. From federal mandates for public awareness for pipeline companies to the ever-constant need to promote 811 to new generations and first-time diggers, the task not only doesn't end, but requires new creative thinking and different engagement strategies on new and evolving platforms.

That makes public relations, communications, journalism, marketing, and other undergraduate majors just as applicable to the damage prevention industry as a certification for locating. It isn't that one of these is more important, but each is a critical part of the overall project of getting every proposed dig to be planned in advance, coordinated through 811 Centers, marked by utilities, and damages avoided.

Part of the storytelling in this industry is to ensure people know it is cutting edge and employing the latest technology and practices. That can go a long way in ensuring confidence in the system. As marketing, awareness, and education professionals do share that message, they have a wealth of innovation to reference. Drone-mounted GPR, AI integration, and software-enhanced visualization tools are promising, but Ron Peterson warns against overhyped gadgets: "Sometimes they work. Sometimes there are so many false positives, it's a guess." That's why it is high-quality, skilled, and well-trained people that are needed as this industry moves forward. Not only to use those gadgets, but to interpret their findings and know what to do next.

Ten Years Ahead: New Talent, New Tech

Envisioning the future 10 years out has always been a challenge and typically only a handful of visionaries can even get close like Steve Jobs and Elon Musk. Some people are sure they have the future figured out, but it doesn't shape up as expected: See Dean Kamen (Segway inventor). Rapid technology advancements and AI will no doubt make the changes over the next 10 years mind boggling, even to the Gen Zs, but taking the time to take a stab at it is worthwhile.

> There's never really been a better time to embrace some of this new technology and innovation in our industry … members that take part in predictive analytics and risk scoring have seen more than 50 percent reduction in damages.
>
> (Kyle VanLandingham, Texas 811)

In conjunction with ACTS and their Global 811 Magazine newsletter, we asked damage prevention industry professionals what they imagined the industry would be like in 10 years. It is illuminating to see the results and creative visions these men and women advance.

Asked what the *best* path to overcoming the biggest hurdles and challenges in the industry in the next 10 years, industry experts replied with the following breakdown:

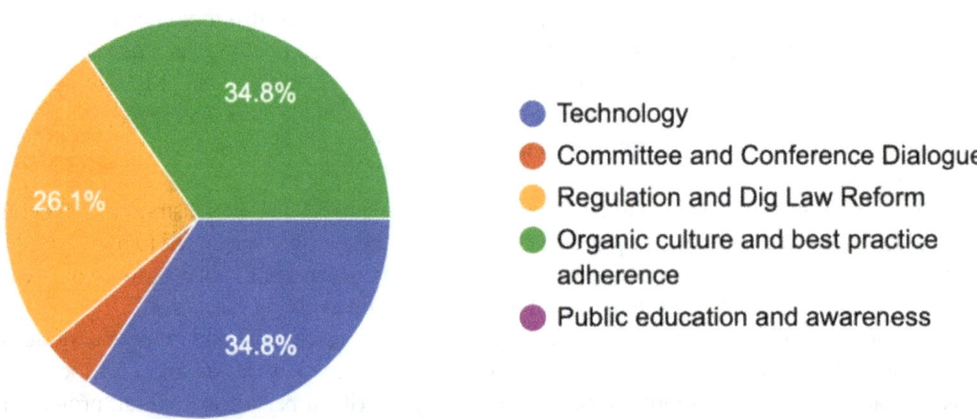

Survey results from industry stakeholders, collected in September 2025.

Unsurprisingly, technology was tied for the top spot. And again unsurprisingly, it's tied with a people-oriented concept of culture. It is clear that the future of damage prevention will be driven by skilled people employing innovative technology.

A summary of highlighted responses for the future of the industry include:

- Having a seamless experience from inception to design to installation and protection.
- The culture of "that's not my problem" has been replaced with "oh, that's not quite right – let me just improve it right now". That appropriately embedded incentives have succeeded in kicking off a flywheel of healthier and valuable culture.
- I hope the industry will be more open to, yet understanding of, technology. I hope the industry has embraced a true healthy skepticism as data is presented and open, respectful, dialogue, driven by questions, with the belief that answers lead to more questions, is the norm.
- Virtual Agents will process all locate requests.
- There will finally be common ground between stakeholders.
- This industry will develop the methodology to break down barriers and resistance to change for the common good when the benefits of speed far outweigh the downsides of "what's really in it for me" hurdles.
- We will finally see national or federal certification program for locators and excavators.
- We'll achieve real trust between all the systems. A system that fixes itself with continual learning and adapting to create a better preventive tactic.
- Excavators lacking full access to the information they need to protect underground infrastructure will be addressed.
- It is going to be critical for the utility locating industry to become a skilled trade. In order for that to happen, utilities need to take it seriously and start paying their in-house or third-party locator what a skilled trades person is worth. That is the only way it will be possible. At that point, we will attract more people who are qualified and we will be able to retain them. The potential will be greater for locates to be completed on time and accurately. This alone would reduce the amount of damage and increase safety tremendously.

While these do not include every response given, they summarize many of the high-level visions for what lies ahead. Central to all of them is the foundational damage prevention mantra of communication and collaboration, put succinctly by one member as:

- Knowledge! Via communication, education, data, and collaboration. When we know better, we do better as an industry.

When asked, many industry veterans spoke about system dynamics, processes, and procedures they expect to improve over time. Many even nodded to or invoked technological advancements as well. But these responses and many interviews we conducted may still not give the reader a full picture of the type of technology just on the cusp of becoming standard in this industry.

Young readers considering a career may want to know about the fast-paced futuristic trends coming up. There is no shortage of technology being developed and making its way into the industry. We have mentioned several already, like fiber optic sensing, but a bit more depth may pique further interest. This still only scratches the surface, so eager readers should take this simple overview as a challenge to create and improve the technology and industry themselves, writing the next chapter of how technology is integrated with their own careers.

There are multiple uses of drones. Small drones have already been in use in the damage prevention industry for years, and their prevalence is only on the rise. Utilizing cameras, drones often serve as a way to conduct surveys, gain an aerial perspective of a dig site, validate or produce new maps, inspect a damage, or identify a leak (including in the offshore and marine context). In its 2021 Technology Report, CGA even explained that it expected that by 2030, drones would be used for the ideal excavation project, integrated with GPR technology and validating the presence and location of utilities in real time to protect excavators.

Not all drones are flying, however, the advent of crawling robots and other land-based unmanned vehicles is expected to further bolster the industry. This will take hardware and software engineers, and visionary entrepreneurs to hone the technology and create new applications for it. With GPR vehicles being able to scan a worksite without a human pushing it, this could not only increase the efficiency of a job, but safety as well, as the locator, excavator, or facility owner could be on site doing other tasks or verifying additional information.

Other integrations of technology on the horizon include improvements to locating and facility identification, including Bluetooth, Radio Frequency Identification (RFID), and even satellite linkage for lines buried underground. Suddenly, the landscape looks far more futuristic than a 20th-century industry of churning up dirt, laying a steel pipe, and using a metal detector-like wand to locate it. Already today, and more so in the near future, the technology in the damage prevention industry will include advanced and sophisticated mapping, bolstered by unmanned vehicles, computers talking to one another from underground, surface level, and in low earth orbit. These will help facilitate better record keeping, which will be continuously improved through further SUE investigations, digitalization, and AI that cleans up, verifies, and creates risk assessments for every inch of the underground.

While some of this technology and the processes above are already in the works, others still need to be fleshed out or refined. Exactly how and when that will happen remains to be seen.

Attracting Talent to Make a Difference

People will be the biggest factor in determining how many of these ideas will come to pass. The industry is full of talented and passionate people, but making these futuristic concepts a reality will be more likely with an influx of new people. All the current research on job satisfaction indicates that people, especially Millennials and Gen Zs, put a high value on working at jobs and companies that make a positive impact on the world. Even though Scott is a Baby Boomer, the mission of the damage prevention industry has motivated him for 40+ years. Looking back, Scott also feels that many of the team members at Rhino and ESA stayed with these companies because of the companies and industries mission to save lives and make excavation safe while protecting our buried infrastructure.

Chris Czarnik, CEO at Career [RE]Search Group, is an expert on hiring and talent management. Chris was the keynote speaker at the 2025 Global Damage Prevention Summit and his topic was extremely timely: *Winning the War for Talent*. Chris pointed out that the bulk of the workforce is always composed of three generations. Right now, the three primary generations are the Boomers, Gen X, and Millennials. Over the next three to five years, the Boomers will all be out of the workforce, which leaves Gen X, Millennials, and Gen Zs to fill the jobs. The catch is that this new group is almost nine million smaller than the current workforce. Even with Gen Alpha soon to enter the workforce, hiring is likely to become more and more difficult as the workforce shrinks.

Tips from Chris on Attracting New Talent to the Industry

There are really three things to keep in mind if you are attempting to connect with late millennials and Gen Z about considering a career in the damage prevention field:

1 Virtually none of them have had any training on how to choose the job or the industry that is a good fit for them.
2 Almost none of them know what damage prevention is or even have an awareness that this industry exists as a possibility for them.
3 The only person who knows who fits in this industry is the employer, not the job seeker.

Let's deal with each of these challenges here.

It is unreasonable to think that a person who has never seen the work that damage prevention does or has experienced it any time in their life will choose it over all of the others. People choose to become firemen because they see firefighters and fire trucks. People choose to become police officers because they see police cars and interactions as part of their daily life. Think of it in these terms: "What did YOU know about this industry before you ran into someone that knew about it?" Likely nothing. *It is most likely that you had never even considered this career until someone explained it to you and told you that you would be a great fit there!* So if that was true for you, then keep that in mind as you are trying to think of ways to influence young people to consider this industry … someone has to tell them that they belong here.

With that in mind, take a look at one of your job ads. Likely it is made up of duties, tasks, responsibilities, and deliverables of the job. Quick question: "What good does it do to describe a job to someone who has never seen it or knows what it is?" For example, if someone has never done or seen welding, what good would it do to tell them the duties, tasks and responsibilities of a welder? They literally have no idea what they are talking about. Also, doing so makes your job ad look like every other job ad on the job boards. But what if there was another way?

If the job seeker doesn't recognize the job, then the only rational thing to do in the job ad is to describe the ideal candidate … who they will recognize as themselves!

The key idea is: "Don't describe the job in the job ad, describe the person that we hope applies to this job."

Since so many people that we need will be at the entry level (we are going to train them in what they need to know) then in the job ad describe the person that will love this job and will find it a natural fit for their skills, values, and interests. You can do so easily by surveying your best entry level employees and creating a persona of the ideal applicant. You can do so by asking them these questions:

1 How do you spend your free time on nights, weekends, and holidays?
2 What are your favorite hobbies?

3 What were your favorite classes in school?
4 What types of side jobs have you done to earn money that you really enjoyed?
5 What types of projects do you find yourself volunteering to help your friends out with?

Now with this information in hand, use the job ad to describe the person who will love this type of work. The job ad might look something like this:

> You knew you were meant for more than fast food … come find more here!
>
> How would we describe our ideal next hire? It's someone that loves camping, fishing and hunting. It's someone who does odd jobs like roofing, drywall of concrete work on the side with their buddies for a little extra cash. It's the person that loved their industrial arts classes but was frustrated by English class. How do we know all this? We asked our best employees to describe themselves.
>
> If this sounds like you, we already know you will love this job … come ask us how we know.

The last tip here is not to focus on the pay and benefits of the job … focus on what that pay will buy for them. Whether it's a four-wheeler or a "crotch rocket" motorcycle, or it your first house, we can get you there. See, people don't choose a career in their early work years … they choose a job that get them the things they see on YouTube.

Don't focus on the job … focus on the person you hope to get the attention of and you will be way ahead of everyone else.

Taking Advantage of Chris' Advice

We believe that as an industry, stakeholders can team up to go after the top talent. The damage prevention and excavation safety industry's mission resonates with the younger generations. When you combine the mission with the ability to make an impact and the broad range of job opportunities, this industry can be a magnet for talent. Now the industry needs to come together with a consistent message for people entering the workforce. The points below are merely a starting point for this discussion.

1 Define the industry mission to create a short tagline or phrase that stakeholders who want to participate can use.
2 Create a consistent way to explain how an individual can make a difference. Examples could include the creation of successful public awareness campaigns developed by individuals or small grass roots groups, like 811 runs, 811 jockeys, 811 hot air balloons, providing 811 tree tags to nurseries, etc. There are countless examples and whoever is doing the hiring could likely showcase examples from their company or stakeholder group.
3 Create a list of all the jobs that fall into the damage prevention and excavation safety industry. Add to this list the trend toward industry certification programs and the picture painted will be compelling.

Tools and messaging may be created that all stakeholders can share, like the NSDM banners created and shared by the 811 Centers. Consistent messaging would help ensure the public sees damage prevention and safe excavation as an industry, and one that keeps the lights on and people safe.

Food for thought: It may also be beneficial to create an industry "job fair" or "industry awareness fair" playbook. The idea is to have these fairs at industry events and invite colleges and/or high schools to participate. The programs could be as simple as creating a template for a presentation for the kids attending followed by allowing them to check out the expo. Depending on the event and the location, there could even be a field trip or demonstrations (locating, GPR, technology demos, etc.) specifically aimed at the youth. Blending a job fair into a conference expo could be an incredible opportunity for exposure – both young talent to the industry and for expo presenters and sponsors to have their brand be the first thing people see.

Permanent Industry, Continuous Improvement

This industry will never disappear, because the pipelines and cables that our countries run on will always exist. In fact, since abandoned infrastructure is rarely removed, and new cables and pipelines are installed daily, the need to protect the infrastructure will continue to increase. The industry has done an amazing job protecting everyone and the buried infrastructure, but technology and new talent will help take it to the next level. As you will see in the next chapter, formalizing a path forward for careers in the industry is gaining momentum.

Chapter 10 Resources

ESA/ACTS Town Halls

Chris Czarnik: Keynote speaker, trainer, coach focused on recruiting and retaining talent.

CGA Technology Reports

A Top 10 list of the real costs of a damage along with information on new technology which integrates GPR into a backhoe bucket.

Note

1 Basques, B. (n.d.). *FuzionView – Sharing 811 (call before you dig) data.* FOSS4G NA 2024. pretalx. https://talks.osgeo.org/foss4g-na-2024/talk/MN7JN8/

Chapter 11

Certifications and Career Mobility

The classroom was not where Jamal was usually found on a Saturday. Normally, he would be at home playing with his three kids, helping his wife around the house, or doing some work in the garage. A classroom was really never where he typically found himself. During the week, Jamal was moving from site to site, kicking dust off his boots, or switching into or out of a reflective vest and hardhat.

A fair share of his time was spent in or between offices and air-conditioned buildings, but he had cut his teeth in the field, and that was where he was most comfortable. But he has grown a lot since starting in this industry. He had spent summers in his teenage years working for his uncle's construction company. After high school, he picked up an associate degree at the community college up the road, paying his tuition with money from odd jobs like landscaping. He had the option to transfer to a four-year college, but honestly found the time outside and in the dirt more fulfilling, so he collected his A.S. and went to work doing contract jobs.

Fortunately for him, contracting led to ample time *in the dirt.* More specifically, he was moving dirt. Jamal became the most proficient dirt mover in the area and gained a word-of-mouth reputation as a reliable and recommended excavator. The contracting job is even where he met his wife! While he was digging out a hole for a client's pool, a neighbor's daughter – home from college – poked her head through the fence. The rest was history!

It was professional contract excavating work that bought them their starter home, put food on the table, and helped them name their third son, "Sinko." Really, they called him "cinco" because he was the fifth member of their family, but it became a family joke to call him *sinko* because he loved to mimic daddy and dig holes with his bare hands and plunge into any mud pit he could find.

Those years brought a lot of joy, a lot of back pain, and a lot of sweat. It was fulfilling, but more than anything, it gave Jamal a sense of a wider world he wanted to conquer. With a family now, his youthful years behind him, and a great deal of knowledge collected over years of digging across the region, he wanted to do more. It was that path that led him to locating.

His wife retorted that he had *traded dust on his boots for paint.* Dust, she could clean off. She was not thrilled about the prospect of multicolored boots in the house. They compromised on the boots staying in the garage, but she was much happier that he wouldn't be down in any trenches or operating heavy equipment anymore.

DOI: 10.1201/9781003796619-13

While locating, he was able to apply a lot of the knowledge he had about excavating and various utilities to his everyday tasks. More than that, he learned something new on virtually every job. The learning and the overlap between his current role and previous jobs also led him to wonder how else he could tie together these silos of utility damage prevention. Surely an industry this robust and critical to public safety had a single thread running through each role, each stakeholder group, each aspect of utility safety that he could pull and tie it all together in a neat bow. That's what he began to seek, not merely to be a great locator, but to be a damage prevention professional.

...

He recently heard about additional career development opportunities that could guarantee upward mobility for him, stability for his family, and interesting work that keeps him engaged and fulfilled while making a meaningful impact for his community. He would be able to master the ecosystem.

If before, he had siloed knowledge – deep but limited in scope – he would now walk the spider web of the entire industry and pull on the strings of each individual discipline as he learned it and moved to the next. Jamal will be a damage prevention professional, a Swiss Army knife of insight, and ready for a role at any organization at the highest levels of strategy, asset protection, and more. And it may be the last certification he ever needs.

Smart, ambitious people like to see career path options when they evaluate opportunities. The damage prevention industry provides people with an opportunity to make a difference, including potentially saving lives. However, the damage prevention professional certification Jamal is pursuing in the chapter introduction does not yet exist in the real world. Right now, there is often no clear picture of the career options available and how to advance from one step to another. There are no certifications required for roles in damage prevention, other than an engineer.

> You look at damage prevention. There are so many roles and responsibilities or even jobs that contribute to preventing damage, but there's no curriculum that somebody can go into and say: I'm going to be a damage prevention professional.
>
> (Mike Sullivan, President of Utility Safety Partners)

Requiring locators to have some type of certification has been a hot topic for as long as we can remember, but it has finally started becoming a reality in Australia, parts of Canada, and through a Nulca certification program available in the United States. This is needed and will help the industry. The catch is that, as a stand-alone certification, where can it take a locator in terms of a career path?

New entrants to damage prevention may not even know how their role specifically relates to each other stakeholder group. Visualizing the entire ecosystem can be complicated on its own, but it is even more complicated to imagine the direct links vertically, horizontally, and slantwise that a person may move in their career. Unless someone wants to start their own business and become an independent contractor, small utility owner, or locator, everyone works for an established company, organization, or government. Seeing the links between all of these and how professional certifications can bolster credentials and hard skills is vital to the longevity of the industry.

The Damage Prevention Professional Certification Program now being developed by Utility Safety Partners in Alberta represents a significant step toward addressing this gap.[1] While its rollout will take time, the very effort to formalize professional standards signals a turning point – one that could improve safety and make the industry more attractive to new talent.

Certifications and Accreditations Exist, but Are Not Widely Required

Training is common in every workplace, yet certifying competence and skillsets goes well beyond onboarding.

> The responsibility lies currently at the company level, because it's the company's responsibility to train their people, because ultimately, it's their liability insurance and the engineers assigning insurance that's up against it if there's an issue that shows up.
>
> (Lawrence Arcand, PE, P. ENG, 4Sight Utility Engineers)

As Lawrence Arcand pointed out, training is the employers responsibility, but the industry has recognized the need for standardization and certifications. The certification programs that exist have been developed by specific stakeholder groups, so they are somewhat narrow and are not part of an overarching program.

The interdisciplinary nature of damage prevention and necessity to communicate and collaborate across roles and stakeholder groups make a multifaceted certification a natural need. Moreover, the stakes in this industry are literally life and death, and billions of dollars of disruption. Other industries routinely require certifications for far lower stakes.

> Well, my hairstylist has to have a certification to cut my hair. We're out here protecting multi-billion dollar facilities that can kill people. And, you know, I can go online, buy a piece of locating equipment tomorrow, call myself a locator.
>
> (Ron Peterson, Executive Director Nulca)

While licensing boards and certifications can serve as gatekeepers, the real goal behind certifying safety professionals is to ensure they have a systemwide view of what their role is and why it matters. Similarly, many professions require continuing education to ensure they are up-to-date with the latest practices and technology. While most knowledge is gained on the job and through years of experience, certifications provide a broad but robust foundation on which to build and apply that on-the-job knowledge.

Perhaps the first certification program was internal to a single company. That company, So-Deep, was founded in 1981, and achieved recognition as being the original subsurface utility engineering company. Their program included all aspects of utility locating, which were detailed and used for training and promotion. At a broader industry level, it took a decade to see a formal standard established.

Nulca, the industry association representing utility locating professionals, established Competency Standards for training utility locators in 1996, and as of 2025, these standards are in their 5th revision.[2] This standard serves as the basis for the Nulca Accreditation/Certification program, rolled out in March 2016 through a partnership with the National Sanitation Foundation (NSF), the industry leader in safety-based risk management solutions and verification.[3]

Accreditation enables locators to provide proof of their consistent training standards and training delivery methodology. Nulca has partnered with NSF International Strategic Registrations (NSF-ISR) in order to provide additional credibility to their Accreditation.[4] NSF-ISR now independently reviews all training materials, processes, and programs related to the Nulca Competence Standard and ensures that members are meeting the requirements set forth in the standard. NSF-ISR was chosen to carry out the independent review based on their experience as a global leader in certification, verification, inspection, and testing services. In order to earn the Nulca Accreditation, an organization's program must be reviewed by NSF-ISR auditors to verify that all 10 components of the Nulca Competence Standard are met.[5]

The standard has been recognized by the CGA Best Practice Guide, which identifies and validates the consensus best practices to enhance safety and prevent damages to underground facilities.[6] This Accreditation program is excellent, but it only applies to locating. Some companies may require locators to have been trained to these standards, and it can be a requirement in some bids and contracts, but it is not an industry mandate. As of 2025, New Hampshire requires locators to be "trained in accordance with" Nulca standards, while locators in Virginia and Connecticut must be trained to an equivalent level to Nulca's program.[7] [8] [9]

The "10 Components of the Nulca Competence Standard"

1. Basic Locating Theory
2. Use of the Transmitter
3. Use of the Receiver
4. Marking Procedures
5. Knowledge of Facilities
6. Visual Observation Skills
7. Safe Work Practices and Regulations
8. One Call Regulation, Requests and Documentation
9. Excavator & Customer Relations
10. Locating Pipelines

Staking University is a private company that has been providing hands-on locator training since 1999.[10] Online video-based training is also available and covers instrument theory, map reading, utility systems, visual observation, map ticket integration, and more. They offer Utility Locator Certification, and they have training classes running from two to five days.[11] Training options for EM, GPR, and a Train-the-Trainer course are available. The Train-the-Trainer program allows a company's in-house trainers to certify locators to Staking U standards.

Staking U has a campus in Manteno, IL (just south of Chicago), which includes live utilities and almost every locating scenario a locator runs into in the field.[12] They also do on-site training for companies and hold workshops around the United States. Staking U has delivered training to over 600 organizations, including utilities, municipalities, engineering firms, contract locators, military bases, and more across the United States, the United Kingdom, and Australia.

Casper College (WY) delivers a utility locator certification program, where graduates receive certification from Staking University.[13] Having colleges and technical schools offer certifications related to damage prevention may become an integral part of a Damage Prevention Professional Certification program.

Across the industry, and certainly in related or peripheral industries, many certification programs exist. One such is the AFA certification, which validates that stakeholders follow the laws, complete trainings, employ best practices, and most importantly can safely install a fence.[14] DIGIN Midwest offers Pipelayer and Erosion and Sediment Control Construction Site Management certifications.[15] Others similarly certify within their specific industry or subsector. They are mostly siloed and particular.

International Examples

Choice of locator can often be an important factor in reducing damages. That is in part because this enables contractors to identify well-trained, certified, or credentialed locators.

> In New Zealand you can pick the utility or your own locator – and the damages went down.
> (*Duane Rodgers, Pelican Corp* Utility Locate Fees A Solution Town Hall)

Utility Safety Partners (formerly Alberta One Call) began an Alternate Locate Provider (ALP) program in 2024.[16] The ALP Program provides excavators with the option to hire an approved locator to complete locates on behalf of participating USP members. The goal of the ALP Program is for project owners to receive locates in line with their project requirements and provide certainty around project timing and costs.[17] The Alternate Locate Service Provider (ALSP) works directly for the project owner on their timeline and can serve to reduce unnecessary downtime to improve project outcomes.

An ALP locate request is submitted through Utility Safety Partners' standard ticket creation process by selecting "YES" in the ALP Option field. Once submitted, the ticket requestor works directly with their selected ALSP to agree on payment and project execution to have locates completed for members participating in the ALP Program. Members who are not participating in the ALP Program will receive a copy of the ticket and will respond to the request following normal procedures.

Alternative Locate Service Providers must be registered to participate in the ALP Program and are required to meet specific criteria defined by infrastructure owners, including specialized training, quality assurance audits, reporting, insurance, experience, and capabilities.[18] USP does not use the term certified, but the training program the ALSP must go through and pass is very rigorous and comprehensive.

In a calendar year between August 2024 and 2025, the ALP Program was utilized 35,678 times. Astonishingly, during this same period, the total number of damage incidents at ALP sites was two.[19]

While the Canadian Association of Pipeline and Utility Locating Contractors (CAPULC) does not directly confer certifications to individual locators, it endorses and supports a competency-based certification framework across Canada. CAPULC created a National Underground Facility Locating and Marking Standard and has developed an Education and Training Partner Program with criteria that must be met to be listed as a partner.[20]

Australia's CERTLOC, founded in 2017 as DBYD Certification and rebranded in 2024, is one of the most structured and formal locator certifications in the world.[21] It is an independent, not-for-profit body that certifies both individual locators and entire organizations, with additional certification for vacuum excavators in development. With more than 1,100 Certified Locators and partnerships across major utilities, CERTLOC is credited with raising industry standards and significantly reducing strikes.[22] Unlike the United States and Canada, CERTLOC certification is increasingly required across Australia, including many state road authorities, local councils, government departments, and major utilities.

According to Chris Ross in an October 19, 2025 statement:

> There is now work on a single national locator qualification, proposed by BYDA and endorsed by BuildSkills, the national jobs and skills council for the construction sector in Australia. This would result in an industry approved unit of competency and targeted training pathways including practical skills assessments. After inputs on scope captured through workshops across industry, technical working groups have developed content which is now released for comment. There's also work on a unit of competency for breaking ground near utilities to be incorporated into other relevant training pathways for excavators.

The Damage Prevention Institute and the Gold Shovel Standard

The Gold Shovel Standard (GSS) was a safety certification program created by Pacific Gas & Electric (PG&E) around 2013–2014.[23] The intent was to ensure that contractors digging around its pipelines were properly trained and accountable, reducing utility damages during excavation. The goal was to provide a credential that asset owners and municipalities could require of their contractors to ensure they followed best practices for safe digging. In concept, this also leveled the playing field for contractors wanting to do work for companies that required the GSS credential.

Effective in 2016, PG&E only hired contractors who were Gold Shovel certified.[24] In the first year of the program, PG&E reported a 36 percent reduction in 2nd party damages.[25] This program was converted to a non-profit organization tasked with rolling out GSS nationally. The program had some success with major utilities like Xcel Energy, conEdison, Dominion, Kinder Morgan, and TELUS signing on. However, GSS did spark some controversy and pushback from some contractors.[26]

In 2023, the CGA acquired the Gold Shovel Association (GSA) and launched the Damage Prevention Institute (DPI).[27] At that point, the GSA was dissolved. The DPI is a program that has been established by CGA to address systemic inefficiencies in the damage prevention process through the development of a comprehensive participant accreditation and elevated metrics, creating the foundation for a consolidated benchmarking and true peer review process.[28] The DPI's objective is to build on the GSA's progress and take the industry to the next level by focusing on measuring the outcomes of all participants in the damage prevention process, while also assessing the systemic impacts on the industry overall.

This may be the closest thing to a full certification program in the United States, but no formal professional certification has been proposed from DPI. Rather, participating members are held to higher standards for damage reporting, communications practices, and other accountability metrics, while signing a pledge to adhere to additional terms and best practices. Still, this could be the natural launching point of a multidisciplinary professional certification that would not only strengthen existing stakeholder reputations but attract new talent into the industry and its many important but often overlooked roles.

Because so many roles exist within the damage prevention industry, as seen further below, it is important that all related stakeholder groups consider the question of certification and how they can participate in a comprehensive program. From the engineering side of things, James Anspach noted that:

> Although maybe not specifically a damage prevention alone practice, the American Society of Civil Engineers (ASCE) is leading an effort for a Professional Engineer (P.E.) to get an additional certification – that of a Certified Project Utility Engineer – which will cover all the engineering requirements including locating, damage prevention, coordination, and so on. Expected to be on-line and available by 2026.

This type of certification, particular to professional engineers, would nevertheless go a long way in establishing a precedent for a comprehensive Damage Prevention Certification (DPC). These individuals could then go on to work in roles across the industry, even if they leave traditional engineering roles to take on more project management or even consulting positions.

Careers in Damage Prevention

Damage prevention requires people in many different disciplines, but they may not be recognized as being part of damage prevention. Changing this perception is all about education and communication. Once the various damage prevention jobs are linked, career paths will be more likely to follow. Below are some of the roles that are part of the damage prevention ecosystem:

- Locate Technician
- Utility Locator Supervisors and Managers
- 811 Center employees
- Public Awareness
- Claims/Damage Investigation
- Excavation Safety Inspectors
- Field Safety Representatives (Utility/Contractor)
- SUE Engineers
- Engineers who design pipeline and cable routes
- Utility Operations/Grid Dispatchers (Central Coordinators)
- Surveyors
- Data Analytics
- Utility Mapping and GIS Specialists
- Right-of-Way
- Damage Prevention Program Managers
- Training and Safety Instructors
- Pipeline Integrity Specialists
- Emergency Response Coordinators
- Regulatory Compliance Officers
- Utility Operations Planners/Schedulers
- Construction Project Managers (utility-focused)
- Utility Risk Analysts
- Business Owner
- Compliance Officer
- Construction Manager
- CEO
- Director of Safety
- Executive Director
- Engineer
- Education Manager
- General Manager
- Marketing Manager
- Public Works Director
- Public Awareness Liaison
- Public Awareness Manager
- Operations Director
- President
- Project Manager
- QA/QS Gas Compliance
- Quality Assurance Manager
- Risk Manager
- Safety Manager Training
- Utility Compliance Manager

Damage prevention is embedded in nearly every utility and infrastructure company, but the breadth of these roles is often hidden or siloed. From locators and right-of-way staff to engineers, dispatchers, and risk analysts, countless professionals are working toward the same mission – protecting people and facilities from damage. Companies should make these roles more prominent and frame them explicitly as part of the damage prevention field. Just as importantly, they should highlight how each role connects with others across sectors, reinforcing damage prevention as a unified, cross-disciplinary profession rather than a series of isolated tasks.

Do you have one of these jobs but never been told you work in "damage prevention?" Do you have a role related to one of these? Come to the next conference or reach out to any person quoted within this book and join the community you already belong to!

This excerpt from Scott helps explain a part of the problem. While explaining the naming of a magazine, he illustrates a key challenge related to identity.

> The magazine launched in 2010 as Damage Prevention Professional magazine. That title was very descriptive of our readers, but it was a mouthful to say, so in 2019 we rebranded to … dp-PRO. There are hundreds of thousands of professionals across the world who have jobs involving damage prevention, but only a small portion of them actually view themselves as damage prevention professionals. This means the name dp-PRO doesn't click with the majority.

In time, the magazine changed names again to better resonate with the industry. The solution required to get every person in this industry to realize they are a professional and to identify with it may simply be to create a new program and label that tells them this truth. Two ongoing initiatives demonstrate this trajectory.

Developing the Next Wave of Industry Leaders

From earning the rank of Eagle Scout (Benjamin) to completing corporate trainings and running a national think tank, I've been exposed to thousands of messages on leadership. The one that has made the most impact, however, came from my church. The lesson is simple: the number one job of a leader is to grow.

It is something the church elders and staff ensured every leader in the church knows and commits to. The reason is that a strong organization needs healthy leaders, and healthy leaders must have humility and commitment to continuous growth.

It is the same principle as the flight attendant mantra to put your own mask on before helping others. If you aren't growing, how can you be expected to lead, steward, or disciple others? My church does this well, if imperfectly, but they are continuously cultivating leaders. Leaders are also asked to pledge to replace themselves by further investing in others who can take the mantle, whether it is leading a bible study, community group, or wider ministry activity.

The model they use is to equip far more people than there are leadership roles, because over time, new roles emerge and positions open, or people move on. The leadership cultivation starts immediately after membership by inviting groups into an eight-week study course known as "Discipleship Track," where members learn how to grow. Many people take *D-Track,* and from there, several are also asked to participate in other courses or trainings. Not everyone goes on to lead a group or ministry team.

Over time, the church fills up with equipped members who are able to sharpen one another and maintain a pervasive culture of commitment to growth due to so many leaders in the body. It isn't top-down leadership, it's leaders in the middle and outward, filling up the body. Some of them go on to lead ministry groups or take on positional leadership roles. By that time, these men and women have already gone through many formal trainings and equipping and have demonstrated commitment to growth and investment within the church and community.

As one can imagine, this lends itself particularly well to mission work and projects involving construction as well! When the group is filled with individuals who know not only servant leadership, but key aspects of organization and team management, there is less miscommunication and smoother work.

Whatever the organization, having many leaders ensures that when positions need to be filled, there are many great options ready to go, and the organization as a whole benefits, regardless of which individual steps in. In fact, to have lower, middle, and upper-level people trained and equipped in leadership principles and steeped in the organization's missions and strategies only makes the whole machine more efficient and effective. This creates an effective leadership pipeline, always in service. It is no surprise that this concept has been proved within the damage prevention industry, and perhaps less surprising to see the source it came from.

811 Notification Centers are the heart and soul of the damage prevention industry, so having strong leaders within those ranks is critical to the industry. Many of the current leaders have been in their positions for many years, even decades, so as they near retirement, there have been concerns about how these shoes will be filled. Running an 811 Center is a very specialized task. Of course, there are plenty of skilled professionals who can do these jobs, but how can the training and transitions be seamless?

Like many challenges in the damage prevention industry, dedicated people always seem to step up and create solutions; Chapter 8 made that clear. In this case, with critical leadership vulnerabilities on the horizon for the industry, two seasoned Call Center leaders put their heads together to create a solution that could only be designed and implemented by selfless visionaries.

Bill Turner, President and CEO of Tennesse811 and Chris Stovall, President and CEO of Texas 811 were talking about the problem one day and decided to do something about it. They created the One Call Executive Apprentice Network (OCEAN), an immersive, in-person program designed to develop the next generation of executive leaders through intensive learning and mentorship.

Within months, these men put together a comprehensive curriculum and an extensive manual for students. Scott attended a presentation explaining the program at the 2025 Global Damage Prevention Summit and was completely blown away with the toughness of the program and the way it is being taught. The 240-hour course is only taught in-person because Bill and Chris believe that the education is far more effective in-person, and it allows the future leaders from around the country to create relationships, which will allow them to share ideas and mentor each other over the years.

The OCEAN program has already graduated two classes, with 12 extremely well-trained future leaders hard at work making a difference. Notification Center Presidents work long hours with a never-ending list of things they can do to make a difference. Bill and Chris made the personal commitment to create, run, and teach the OCEAN program, along with doing their regular jobs. Kudos to both of their Boards for allowing them to invest the time and resources into a program that benefits the entire industry, not just their states.

What Chris and Bill did is extremely impressive. This kind of initiative and passion is what drives this industry. OCEAN is one example of how individuals in this industry see a need, create a solution, and run with it. OCEAN is not a certification program, but the impact on the industry is creating a tidal wave of knowledge. As more leaders graduate from the program, there will be more trained, equipped, and ready leaders than there are Notification Center executive openings. This can only benefit the industry as these men and women take other or supporting roles alongside those specific key leaders. And as the industry becomes more professionalized and cohesive, these OCEAN graduates may take on industry-wide damage prevention roles, which forthcoming certifications may have a role in shaping.

A Damage Prevention Professional Certification Program

The mission of the damage prevention industry is to save lives and protect vital buried and aerial infrastructure. The industry offers all types of careers ranging from locating to engineering to marketing and everything in between. This mission combined with the variety of opportunities should have a magnetic pull as a career, but we are missing clear career path options. The USP DPC Program is intended to solve this problem. It will be rolled out in Alberta, but it is being designed to work everywhere, with localized modification. This idea has moved from a concept to an initial launch planned for 2026 in Alberta.[29] USP has made the commitment to hire a leader to lead the development and launch the program.

The program will be a formal designation recognizing individuals as Damage Prevention Professionals. The term "designation" may be substituted with "certification" depending on the country it is being implemented in. It is being structured like other professional certifications (e.g., CSP, ASP, NCSO), with multiple levels of achievement based on both education and experience. There will be a core curriculum plus specializations/electives. The core curriculum will cover all the fundamental aspects of damage prevention and the electives will provide the opportunity to obtain in depth knowledge on specific subjects.

The Core Curriculum subjects are still under development, but the basic outline is in place. Scott attended the two-part workshop on this program, led by Debbie Shelley and Jeff Mulligan, who co-chair the committee working on the program, at the USP 40th Anniversary conference in Banff in 2024. Both sessions were packed with engaged damage prevention professionals brainstorming on topics – both core and elective – to cover. Below is a list of the current core topics (September 2025).

1 Damage Prevention Fundamentals

   - History and purpose of damage prevention.
   - Understanding the "five steps to safe excavation."
   - Roles and responsibilities of excavators, locators, facility owners, and regulators.

2 Locating & Marking Basics – Not intended to make someone a certified locator, but enough knowledge to understand the process.

   - Principles of electromagnetic locating.
   - Reading and interpreting utility maps/tickets.

- Limitations of locating technologies.
- Tolerance zones and proper marking standards (e.g., APWA color code/Canadian equivalents).

3 Ground Disturbance Principles

- Safe excavation practices (hand exposure, potholing, daylighting).
- Trenching and excavation safety basics.
- Ground disturbance coordination and permitting processes.
- Overhead hazard awareness.

4 Regulatory & Legal Framework

- One Call requirements (811/One Call/USP equivalents).
- Provincial/state/federal regulations related to excavation and buried infrastructure.
- Due diligence and regulatory expectations.

5 Risk Management & Incident Prevention

- Root cause analysis of damages.
- Risk assessment and hazard recognition.
- Case studies of incidents and lessons learned.

6 Communication & Coordination

- How to coordinate between excavators, locators, and facility owners.
- Positive response systems.
- Documentation and reporting.
- Public awareness and stakeholder engagement.

7 Ethics & Professional Responsibility

- Professional accountability and integrity.
- Recognition of DPC as a portable, personal credential.
- Continuing education requirements.

Like any effective professional certification, there will be ongoing CEU requirements to keep certifications current. The DPC program will be overseen by USP with support from a DPC Advisory Board.

We believe that this DPC program can have a massive positive impact on damage prevention. This program has the potential to completely change the course of the industry. Once there are credentialed Damage Prevention Professionals, stakeholders are far more likely to recognize this as a necessary department and a profession. Peer influence will mount when stakeholders begin seeing fellow stakeholders staffed with DPCs.

With the DPC applying to all aspects of damage prevention, it will provide credibility for every contributor in the process ranging from locators to engineers and public awareness teams. There will be teams of DPCs, and metrics will follow. Right now, the primary metrics are all centered on damages. With an industry full of DPCs who are focused on preventing the damages before they happen, metrics will form around prevention. As the saying goes, *what gets measured gets done.*

There is plenty of work to be done before the DPC program becomes reality, and it will take years to become an integral part of the industry, but the train sure seems to have left the station. There have been plenty of great things done to improve damage prevention by many committed people and stakeholders, but this comprehensive certification may lead to a long-term seismic shift and improvement in damage prevention unlike many before it.

Chapter 11 Resources

ESA/ACTS Town Halls

Nulca Accreditation and Training

Staking University provide a wide variety of locator training programs

Utility Training Academy provide locator training classes

CERTLOC Certifications is an Australia based program

CAPULC Training is a Canadian based training program

ALSP program created by Utility Safety Partners specifically for Alberta

Damage Prevention Institute

Notes

1 Utility Safety Partners. (2024). Training standards committee. https://utilitysafety.ca/committees/training-standards-committee/
2 Nulca. (2025). About Nulca.
3 PelicanCorp. (2021, June 17). PelicanCorp joins Nulca to enhance the protection of underground assets. Newswire. https://www.newswire.com/news/pelicancorp-joins-nulca-to-enhance-the-protection-of-underground-assets-21416972
4 Nulca. (2025). Accreditation. https://nulca.org/accreditation/
5 Nulca. (2025). Accreditation. https://nulca.org/accreditation/
6 Common Ground Alliance (CGA). (2026, February). CGA Best Practices Version 22.0; 4.5 locator training. CGA Best Practices. https://bestpractices.commongroundalliance.com/4-Locating-and-Marking/405-Locator-Training
7 New Hampshire Department of Energy. (2024). Chapter En 800: Underground utility damage prevention program. New Hampshire Code of Administrative Rules. https://gc.nh.gov/rules/state_agencies/en800.html
8 Virginia General Assembly. (2026). § 56-265.19. Duties of operator; regulations. Code of Virginia. https://law.lis.virginia.gov/vacode/title56/chapter10.3/section56-265.19/
9 Cornell Law School, Legal Information Institute. (2025). Conn. Agencies Regs. § 16-345-3: Responsibilities of public utilities. https://www.law.cornell.edu/regulations/connecticut/Regs-Conn-State-Agencies-SS-16-345-3
10 Staking University. (2025). Staking University. https://www.stakinguniversity.com/about/
11 Staking University. (2026, February 26). Staking University Five Day Flagship Campus Training. https://www.stakinguniversity.com/classes/5-day-main-campus-classes/
12 Staking University. (2026, February 23). Location and Directions. https://www.stakinguniversity.com/about/our-location/

13 Casper College. (2026, March 5). Utility locator. https://www.caspercollege.edu/program/utility-locator/

14 American Fence Association. (2025). Fence installation school. https://www.americanfenceassociation.com/schools/fence_installation_school/

15 DIGIN Midwest. (2026). Pipelayers certification. https://www.diginmidwest.org/general/custom.asp?page=pipelayers_certif

16 Sullivan, M. (2024, July). Statement from Utility Safety Partners regarding the Alternate Locate Provider Program and TELUS' Locate Policy. https://utilitysafety.ca/wp-content/uploads/2024/07/USP-STMT-re_ALP_TELUS_Locating.Policy.pdf

17 Utility Safety Partners. (2024, February). Alternate Locate Provider Program. https://utilitysafety.ca/wp-content/uploads/2024/02/ALP-Locate-Request-Process-Final-v2.pdf

18 Utility Safety Partners. (2026, February 24). Alternate locate provider (ALP). https://utilitysafety.ca/wheres-the-line/alternate-locate-provider-alp/

19 Utility Safety Partners. (2025, October). Do you want to learn more about Utility Safety Partners Alternate Locate Provider Program? Click here. LinkedIn. https://www.linkedin.com/feed/update/urn:li:activity:7369830877607952384/

20 CAPULC. (2024). Underground Facility Locating and Marking Standard (Version 1.0). https://www.capulc.ca/Standard

21 Row, R., & Newman, P. (2018, July 24). DBYD certified locator: A game changer for industry in Australia. https://actsnowinc.com/global-811-magazine/dbyd-certified-locator-a-game-changer-for-industry-in-australia

22 Row, R., & Newman, P. (2018, July 24). DBYD certified locator: A game changer for industry in Australia. https://actsnowinc.com/global-811-magazine/dbyd-certified-locator-a-game-changer-for-industry-in-australia

23 Pacific Gas & Electric (PG&E). (2014, August 11). PG&E's Gold Shovel standard paves the way for safe excavation. PR Newswire: https://www.prnewswire.com/news-releases/pges-gold-shovel-standard-paves-the-way-for-safe-excavation-270818141.html

24 Patni, S. (2015, October 27). Leak Abatement Best Practices Workshop: Implementing Best Practices. https://www.cpuc.ca.gov/-/media/cpuc-website/divisions/safety-policy-division/documents/6pge_leakabatementworkshop.pdf

25 Wells, A. (2016, June). Gold Shovel Standard Program. EPA. https://www.epa.gov/sites/default/files/2017-06/documents/gold_shovel_standard.pdf

26 Trenchless Technology. (2017, June 12). DCA: Concerns remain about Gold Shovel Standard. https://trenchlesstechnology.com/dca-concerns-remain-gold-shovel-standard/

27 Common Ground Alliance (CGA). (2023, January 3). Common Ground Alliance Acquires Gold Shovel Association, Welcomes New Participants to the Damage Prevention Institute. https://commongroundalliance.com/Publications-Media/Press-Releases/Press-Release/common-ground-alliance-acquires-gold-shovel-association-welcomes-new-participants-to-the-damage-prevention-institute

28 Common Ground Alliance (CGA). (2023, January 3). Common Ground Alliance Acquires Gold Shovel Association, Welcomes New Participants to the Damage Prevention Institute. https://commongroundalliance.com/Publications-Media/Press-Releases/Press-Release/common-ground-alliance-acquires-gold-shovel-association-welcomes-new-participants-to-the-damage-prevention-institute

29 ACTS. (2025). Global 811 Magazine 2025 Issue 1. https://actsnowinc.com/global811magazine/archive

Chapter 12

Policy, Government, and Oversight

Life near the Four Corners was good. Rodney had usually stuck around his own stomping grounds, but popped over the state lines a few times a month for this and that. But that was changing with his new entrepreneurial spirit.

He'd just inherited some equipment and a shop from his great uncle's passing. Now in his mid-30s, he was ready to stop the odd jobs and paycheck-to-paycheck routine and to start something he could grow with and establish himself.

While a buddy was over, and with a six pack half finished, the conversation did its usual meandering with some venting, some nostalgia, and some aspirations. Placing his latest empty on a wall of cans on the rail of his porch, Rodney sneezed, causing the entire edifice to collapse in a metallic clattering. The result of this was to create a clear line of sight for his buddy over to the side yard where a mini excavator sat parked next to Rodney's pick up.

"Rod, what do you have there?" he asked quickly, not having seen it when he arrived because he let himself in through the back door.

"You remember Uncle Jay? That's one of his *little beasts* he always talked about. I got a couple of 'em and his whole shop last month."

"Shoot, sorry man, I forgot you said he passed a few months ago. What are you gonna do with it all?"

Rodney was quick with his reply – as though he had the plan all worked out and had known his whole life, even though this was his first utterance – "I'm going to start a contracting business."

This was a great idea. Rodney was well positioned, had his equipment free and clear, and was able bodied and ambitious. Most importantly, he was humble. Rodney knew he didn't know everything, and even though for a lot of years he just did enough to get by, he had a good head about him. He knew he needed to do more research, make a business plan, advertise himself, and maybe even file some paperwork with the county or state.

"If you plan to use those excavators, it sounds like the contracts you have will be trenching and dig outs," his friend chimed in. "Better google 811 and figure out if you need any kind of special license."

Rodney had heard of 811 but had never thought it applied to him – he wasn't doing pipeline work. All the same, he took the advice. The next day, he sat down at his kitchen table with his six-year-old laptop and started searching. After a few different searches for LLCs and contractor licenses, he dove deep in the 811 topic.

Suddenly, his stomach sank. His head went fuzzy, and his blood pressure shifted. He was filled with uncertainty and felt as though it may be hopeless. He wasn't sure he could start this business after all.

DOI: 10.1201/9781003796619-14

One state required two working days of advanced notice, another said two working days but mentioned exclusions for state holidays as well. The third state required three days of advanced notice, and the fourth was back to two days but excludes the day the notice is given. He worried he couldn't keep these all straight and feared that if he had quick jobs come up, he may lose the opportunity to get the work if he couldn't comply with the state rules or ensure he was in compliance. To make matters worse, the advance notice was only scratching the surface – ticket life, tolerance zones, seemingly every aspect of the process varied only slightly between each state.

Living near the intersection of four states had never been an issue before. In fact, given that his driver's license was good in every state because of full faith and credit from the constitution, he assumed his contractor license and ability to work may be the same. Suddenly, he was confronting four specific and different state dig laws, with no national law or overarching framework.

When he did look at the federal rules, he only had more concerns – *federal enforcement*. Now he feared that if he did something wrong in one state, not only would that state go after him, but the feds might as well.

He needed to go for a run to clear his head. While on the road, he saw a father and son duo installing a new air conditioner unit up the road; old Mr. Jenkins on his riding lawnmower; he caught a glimpse of three young guys reroofing a house. The longer he ran, the more he saw – men and women building things, fixing things, making things work. Something burned inside of him, and he wanted to be a part of it.

At the end of his run, Rodney regained his optimism and felt that four states provided four times the opportunity! He hopped in the shower and sat back down at the table. He had more confidence in his plan now, but he was cognizant that it also meant a lot of different laws and policy considerations. No matter, he would simply familiarize himself with them, ensure he is in compliance, and do his work to the highest standard just like he was raised to do.

While 811 has become ubiquitous, it is far from the ultimate unifying force many expect it to be. Like calling 911, no matter where you are in the United States, your call is routed to the relevant state, local, or regional Facility Notification Center. In this way, it is a unified national strategy. It's also a nationwide requirement to call 811 before any digging project.

> The US has been somewhat of a juggernaut in terms of changing that paradigm shift and having that brand, the 811 brand, which really was a game changer. So there's been an evolving story for damage prevention in the U.S. predominantly.
>
> (Mike Sullivan, President, Utility Safety Partners, *ESA Town Hall*)

But you won't find a federal law written by Congress and signed by the president requiring that excavators contact 811. You won't even find a regulatory rule by a federal agency that has nationwide jurisdiction. In fact, the further you go, the more fragmented it all seems. No singular dig law exists, but there are 50 (or more) dig laws for every state, Chicago, Washington, D.C., and even territories.

At the federal level, the Pipeline and Hazardous Materials Safety Administration (PHMSA) has made state funding contingent on state laws requiring calling 811. But it is not just states and PHMSA at play. The Federal Communication Commission (FCC) played the most fundamental role in designating 811 as the nationwide number. Prior to 2005, PHMSA had, in effect, only required states to have a pipeline safety program to qualify for grants. As time has gone on, PHMSA has tightened its language.

A light summary of some recent policy provides a useful introduction:

In 1998, Congress passed the Pipeline Safety Reauthorization Act of 1988 directed "the Secretary of Transportation to require each State to adopt a One-Call damage prevention program for the establishment, operation, and enforcement of One-Call notification systems."[1]

In 1999, the Common Ground Study was published with sponsorship from the U.S. Department of Transportation.[2]

In 2000, the Common Ground Alliance was formed.[3]

In 2002, Congress passed the Pipeline Safety Improvement Act of 2002, which directed the Federal Communications Commission to set aside a three-digit number for the purpose of damage prevention to pipelines and other critical infrastructure.[4]

In 2005, the Federal Communication Commission formally designated "811" as the nationwide number for excavation and damage prevention.[5]

In 2007, the federally mandated effective date of 811 came, marking complete nationwide adoption and use of 811.[6]

In 2011, Congress passed the Pipeline Safety, Regulatory Certainty, and Job Creation Act of 2011, which amended Title 49 of the United States Code (49 U.S.C. § 6103(a)), further tightening the requirements for states One-Call programs to be grant eligible.[7] Full participation from facility owners and increased participation from excavators would be required to receive grant funding.

In 2015, PHMSA issued a final rule establishing criteria to determine the adequacy of state excavation damage prevention enforcement programs.[8]

> If a state doesn't use effective enforcement as one of their tools in the toolbox to mitigate excavation damages, then the federal government can ultimately have backstop authority to go in and conduct that enforcement in those states. That led to a rulemaking in 2015 that gave PHMSA that backstop authority to, again, conduct enforcement against excavators in states that are ultimately deemed inadequate, meaning their enforcement programs are inadequate to serve the intent of the [2006 PIPES] act.
>
> (David Appelbaum, Senior Transportation Specialist, PHSMA)

An inadequate rating from PHMSA typically means a state dig law has penalties or consequences for failing to notify 811 or otherwise follow safety procedures around pipelines, but no centralized agency or law enforcement authority is bringing those risky workers to justice. That may simply mean not issuing fines or failing to follow through when an excavator damages a line. This type of "inadequacy" enables PHMSA to enforce the state's own law by fining the contractors directly.

So today, we may hear, "call 811, it's the law" and it's true, but only at the state level. From the federal standpoint, it was a carrot and stick approach to lead states through grant funding to make it law. This is the case with many policies in the damage prevention ecosystem. Federally, there are certain guidelines, best practices, grants, and recommendations, but few actual rules and regulations.

PHMSA was instrumental in achieving the implementation of 811, working not only with the FCC but the various state One-Call Centers and state-level damage prevention authorities. The man overseeing that implementation between 2005 and 2007 was Brigham McCown. Serving as the first acting-administrator and the first deputy administrator of the new agency, McCown had a huge task before him.

> It was really one of my highlights of my time at PHMSA – being able to roll out this 811 system. Before then, you had to call different 800 numbers, different local numbers; you could never figure out who to call. So by creating this national one-call 811 number, we wanted to make it very easy for anyone to call one-call.

McCown emphasized the scale of the task in testimony to Congress stating the agency's work "to implement our newest, most important tool is the three-digit dialing for the one-call system" and adding that "it is a big task and we need help to succeed."[9] This underscores the mantra in damage prevention that *communication and collaboration* are key. Needing help to succeed is the very core of damage prevention – true for the top agency and true for the stakeholders on the ground.

Focused on continuous improvement, McCown also saw the vision as one to advance alongside other industries and processes.

> While today, we are advancing damage protection with this state-of-the-art system, it is one I would thoroughly expect to evolve with future advances in technology and innovation.

While McCown went on to run pipeline companies, work in transportation safety, and serve on federal boards, he also founded the Alliance for Innovation and Infrastructure to help ensure robust damage prevention education and research could continue from within the industry and from outside of it – informed by public policy realities. One of those realities is that the federal approach to damage prevention is made up of high-quality but limited jurisdiction directives and incentives.

Consider that PHMSA is the chief federal agency for damage prevention. No other federal agency has as full jurisdiction over the infrastructure and practices of damage prevention directly. Yet PHMSA primarily governs interstate pipelines, which remains just a tiny fraction of all underground utilities.[10] While the agency has expanded its own jurisdiction to add hundreds of thousands of miles of upstream gathering lines to its purview, for damage prevention, its jurisdiction largely stops at pipelines (with its primary domain over hazmat movements by land, sea, and air).[11]

While the U.S. boasts a globe-leading 3.4 million miles of federally regulated pipelines[12] essential to its economy, hazardous liquid and gas pipelines make up less than 10 percent of all underground utilities, and a smaller share of what is damaged every year. In effect, the chief federal agency overseeing damage prevention does not have regulatory reach to 90 percent or more of what is underground. Non-hazardous or local distribution pipelines, water lines, electrical wires, cable and fiber optic lines, sewer and waste pipe, and many other pipes, cables, and wires of different size, shape, and purposes are all outside of federal jurisdiction. These are not governed by PHMSA and are therefore not subject to its direct dictates for damage prevention.

What PHMSA has done effectively is use its leadership and influence to create secondary effects that act as an umbrella over the rest of damage prevention. These are sometimes referred to as knock-on effects. When PHMSA required each state to designate a One-Call Center for pipelines safety, the same center and phone number enabled the protection of electric, cable, water, and other facilities, even without PHMSA having primary jurisdiction over them. When PHMSA issues a new grant for education, call centers and other stakeholders can use it for pipeline safety and conduct outreach or other work that benefits all stakeholder groups.[13]

In other words, whatever PHMSA does directly for protection of pipelines, it indirectly improves protection for all other buried facilities on the damage prevention front.

Central to its non-regulatory contribution to the damage prevention industry is the Nine Elements of Effective Damage Prevention:

1 Enhanced communication between operators and excavators.
2 Fostering support and partnership of all stakeholders.
3 Operator's use of performance measures for locators.
4 Partnership in employee training.
5 Partnership in public education.
6 Enforcement agencies' role to help resolve issues.
7 Fair and consistent enforcement of the law.
8 Use of technology to improve the locating process.
9 Data analysis to continually improve program effectiveness.

This serves as the definitive list of how to craft and maintain a program that protects critical infrastructure and workers.[14] By simply promulgating this educational resource, recommendation, and guidance, PHMSA advances damage prevention and provides a toolkit and model for states to adopt, integrate, or build from.

While for all intents and purposes being the chief damage prevention authority at the national level, PHMSA is not the only federal agency that intersects with damage prevention. The Occupational Safety and Health Administration (OSHA) creates guidelines and binding rules on workplace and jobsite safety. Whether it is hard hat regulations or trench support standards, virtually every excavation in some way complies with a federal dictate, even if one would not naturally call OSHA a damage prevention czar. Perhaps it is more fitting to associate PHMSA primarily with *damage prevention* and OSHA with *excavator safety*, working the problem nationally from different angles.

> When we do OSHA training, we talk about OSHA regulations that are all written in blood. They're there because something bad happened.
>
> (Ron Peterson, NULCA. *ESA Town Hall*)

We have already traced the key contribution of the FCC in designating and maintaining 811 as a nationwide phone number. Yet the contribution of this agency was not a one-time action. In fact, their continued work helps facilitate damage prevention in a more macro sense as the federal agency tasked with regulating and ensuring the efficient operation of phone and communication services nationally. The entire premise of the one-call system relies on a safe, efficient, and secure communication system, including the shift toward fiber optic infrastructure. It may even surprise some to learn of the FCC that "its responsibilities also include the use of communications for promoting safety of life and property ..." which does strike the same chord as the damage prevention message.

Another noteworthy agency, although perhaps more reactive than proactive, is the National Transportation Safety Board (NTSB). As the nation's premier independent transportation accident investigative body, the NTSB regularly visits the site of pipeline ruptures and explosions across the United States, produces conclusive reports, and enumerates pointed recommendations to other agencies and private sector stakeholders.[15] Even when reacting to incidents, NTSB reports serve as highly credible, detail-oriented guides to prevent future damage. By incorporating their recommendations, future incidents are less likely to occur. And while not acting with the force of rule or law, for years NTSB also produced "Most Wanted" guides and recommendations for policy and practices they believe are needed, though this was retired in favor of more informal recommendations in 2023.[16] This proactive set of recommendations serves as another voice joining a chorus of independent organizations advising on best practices and ideal policy for damage prevention at a federal level.

All this top-down rulemaking, guidance, and recommendation helps create a canopy for the country that damage prevention is important and certain standards are expected, while reinforcing that at its core, damage prevention is multidisciplinary. Just as agencies from the Departments of Labor, Transportation, and multiple independent agencies speak into and help provide damage prevention guidance, so in the daily undertaking of damage prevention, multiple stakeholders from excavators to locators, 811 Centers to utility companies must communicate and collaborate.

> The only way the public can have that voice in these particular matters is through the processes that the state has in place, so that those representing the public, which are your state governments, or your one call boards, or stakeholder groups, have the opportunity to take a look at that and weigh in on whatever that outcome looks like.
>
> (David Appelbaum, Senior Transportation Specialist at PHSMA, *ACTS Town Hall*)

If the federal government has so little specific jurisdiction over damage prevention, how does it look at the state level?

State Law

At the state level, damage prevention legislation is often referred to as the "dig laws." These are usually laws passed by the state legislature and signed by the governor, with many additional implementing regulations and rules promulgated by a state agency. Here again, there is an interplay with PHMSA, as each state is required to have an agency or effective enforcement authority that serves this purpose in order to be grant-eligible and listed as *adequate* by PHMSA.

The dig laws vary considerably by state. Some are long and robust, and others occupy only a few short pages. These generally include definitions for common terms like excavation, positive response, and notification center. They outline the basic process of excavators being required to provide notice of planned excavation, utility companies being required to have their records on file and to participate in the call center system, utility locating being required within a certain timeframe – and detailing how long a ticket and markings are valid – what to do if a damage occurs, and more. These basic elements are found in all state dig laws, but a core similarity does not make them all truly alike.

Significant variations in ticket life, notice requirements, ticket size, penalties, enforcement, and more all make different states stand out from one another, in both good and debated ways. Certain voices have called for years for standardization across the states, and even a "national dig law," while others point to the importance of maintaining state laws that are unique and effective in different circumstances.[17,18] Whether any elements are unified and made into a national standard or not, the differences are important to highlight. It is not our task here to advocate for either nationalization or federalism.

The current model strikes a workable balance, with the federal government requiring certain basic elements but allowing the states to refine and implement them in their own way. Moreover, one of the primary functions a national law would have is already in place – federal enforcement.

> We perform the annual evaluations of each state's enforcement adequacy. So, there is a direct responsibility each state has to meet that federal obligation.
>
> (David Appelbaum, Senior Transportation Specialist at PHSMA. *ACTS Town Hall*)

Enforcement

Some may wonder what enforcement means and how it operates in the damage prevention industry. The most basic form of enforcement is what it sounds like, bringing to bear the consequences described when a law is violated. Enforcing dig laws means fining or penalizing a stakeholder who did not follow the dig laws and who caused damage. Enforcement is separate from liability, civil penalties that may arise for contract breach, city fines, and other expenses. Enforcement refers specifically to the provisions outlined in the dig law. Enforcement provisions can include specifically detailed fines, loss of license, and even mandatory training.[19]

> The word enforcement always sort of creates a running away from it. No one really wants to put their neck out there and say that we want additional enforcement or that we need oversight and all these things. But I think the message here that we need to walk away with is that enforcement … is more about accountability.
>
> (Shane Alexander, Director of Damage Prevention, Centerpoint Energy, *ACTS Town Hall*)

Enforcement is a state responsibility. The state writes its own rules and laws, in which they describe the consequences of breaking the law. So not only does each state law look a little different, but their enforcement practices are equally varied. Some states are swift to enforce their laws and keep stakeholders accountable. Others do not fine or penalize violators and enforce their dig laws consistently, and in those cases, PHMSA has the ability to exercise direct jurisdiction and apply federal enforcement authority. In this way, PHMSA is not enforcing the state's law, but ensuring compliance with federal requirements when state enforcement is deemed inadequate.

> PHMSA has a set of guidelines that they use to determine adequacy in the 811 laws, specifically as they relate to enforcement. I think enforcement is a component of this that can't really be overlooked. There's a variety of flavors of it in the country. Some of it's more effective than others.
>
> (Louis Panzer, Executive Director NC 811, *ESA Town Hall*)

In some jurisdictions, enforcement power stems less from fines and more from control over access to the public right-of-way. Cities like Chicago exemplify this model: utilities and contractors must comply with local requirements to gain permits, giving the city a uniquely strong enforcement mechanism compared to many states.[20]

While enforcement has a relatively narrow focus, the broader diversity of dig laws demonstrates how states can calibrate rules to their liking. When evaluating what is most effective, states have an opportunity to learn from one another, and over time, to refine their rules in light of what has worked or what has failed in their own state and others.

Tickets

A common focal point across all states is the waiting period for an excavation to begin. This is typically two to three business days from the time an excavator notifies 811. In many cases, the clock starts at the beginning of the next day, with some states ranging up to 10 days for remote locates, such as in Alaska.[21]

There is an inherent tension here between stakeholders, and public policy is a central element. It can either stoke the tension or resolve it, but it must be grappled with. Consider an excavator – they have a contract to dig somewhere with a paying customer who wants a trench for utilities, swimming pool for a resort, or foundation for a building. Whatever it is, the project owner has a timeline and wants to stay on budget and on time – or ahead if they can get there. Unfortunately for them, the responsible contractor has to provide notice to 811 and wait two to three days depending on state law before he can begin work breaking ground. If well-planned in advance, this is simply built into the project timeline, but issues can arise all throughout the process that delay project checkpoints. Weather, remote locations, and other issues can delay locators from fully marking the site by a day or more, meaning the work cannot begin.

So the excavator wants to get to work. In some cases, he may even have tighter turnarounds. Imagine his customer is not a corporate resort or major infrastructure owner who has planned meticulously, but a local business or individual. That project owner may want the digging to happen as soon as possible and want the contractor to get to work almost immediately. The contractor, then, is under immense pressure: *take this job, get to work immediately, and get paid or wait three days and possibly miss getting the job.*

Cutting corners is not uncommon. In fact, "no notice" is one of the leading reasons for damage. Many people do not contact 811. While there are various reasons for not contacting 811, including not being aware of the process and requirement or not believing what they are doing is considered "excavation," there is ample data to suggest that a sizable proportion of "no notice" damages are committed by excavators who knew full well about 811 and simply proceeded anyway.[22] This may be because they believed they were familiar with the area, they may have had a previous ticket in the same location that expired, they may only be doing shallow work, they may believe they could go carefully and pothole along the way. There may be dozens of legitimate and reasonable justifications for digging without notice (not legally justifiable by the way). But there are also likely many who simply calculated that they could get the job and get paid.

Whether they did the math in their head or not, it's a simple risk-reward calculation. If they believe their chance of hitting a facility is below, say 10 percent, and know they'll bank several thousand dollars if they get the job, they may simply believe it's worth it.

So the excavators may see a 48 or 72-hour ticket timeline as a burden. To them, prolonging the start of an excavation would be unjustified, interfere with their livelihoods, and possibly even incentivize further corner cutting.

State dig laws often carve out exceptions for emergency tickets, allowing excavation to proceed more quickly when life, health, or critical service is at risk.[23] These provisions recognize the practical limits of rigid timelines, though they also raise questions about how broadly "emergency" should be defined and how abuse can be prevented. When "emergencies" take a break for the weekend, for instance, it seems like a clear indication that many are using this as a workaround.

Here again, public policy – and in particular enforcement – has a role in shaping behavior:

> With our Pipeline Safety Bureau over here, they started fining us [contractors] if we don't have a ticket, if we don't follow the law, they fine us. And they started doing the same for the utility companies. And guess what? The accountability started coming up.
>
> (Joel Miller, Spear D Construction, *ACTS Town Hall*)

Now consider a locator. He has to visit the site of every ticket request that is not screened out in advance and identify and mark all the facilities there. For him, it is a race against the clock. This locator may only have 48 hours to arrive on scene, set up, locate, and mark facilities across multiple city blocks. Consider also that ticket requests can be imprecise or large, that Facility Notification Centers add a buffer area, and that many tickets requested may be for planned work later in a project timeline. These together mean a locator may be called out to locate what should essentially be a single lot and have to locate facilities in a whole neighborhood. For roadwork, it may be the difference between locating for 100 feet and locating for nearly a mile or more.

An average walking pace will take a person about 20 minutes to walk a mile. Add in locating and marking, and that mile becomes hours. Add multiple miles, weather conditions, traffic, travel time, and potentially other jobs, and the 48-hour clock begins to look impossible to comply with. And one size is not the same in every location.

Most states have the same 48- to 72-hour clock for all jobs. It's simply a damage-prevention-wide standard. But it means the same two-day clock is running for a locate in a single suburban residential lot as for a major urban road project. As Jason L. Smith has put it:

> One inch of locates?
>
> Two business days ...
>
> One foot of locates?
>
> Two business days ...
>
> One mile?
>
> Two business days ...
>
> 1,000,000 feet?
>
> Still ... two business days ...
>
> 1,000 miles across five counties?
>
> You guessed it – TWO, BUSINESS DAYS.

The opportunity to resolve and refine how the system works through both public policy and industry practice is clear. From these two perspectives alone, there is an inherent tension between excavators and locators – not like movie drama rivalries between the FBI and the

CIA fighting turf wars, but a legitimate policy-initiated tension, where each is trying to accomplish their own work, but working against a legislative clock that doesn't flex to meet their circumstances.

As Common Ground Alliance (CGA) research has explained, "Data from seven states shows that as many as 56 percent of tickets receive late or no positive response."[24] That means late locates or job sites not ready for excavation despite the legally prescribed 48 to 72 hours passing. It is clear there are more dynamics going on across the country than just heavy workloads.

> We need to figure out the root cause, not just give an extra day for locates.
>
> (Roger Cox)

Ongoing research into excavation readiness – or whether a job site is marked and ready to dig in compliance with law after the prescribed timeframe – is working to identify additional factors at play. Not only does CGA do this through its annual DIRT Report and other resources, but Virginia 811 recently published a model, building on Georgia 811's work on this topic.[25] This research explores the intersection of excavation readiness and locator workloads, alongside variables like rural broadband build outs and more. Central to whether a site is ready is the state-mandated positive response.

Communication and Readiness

Required by most states' dig laws, a positive response is the communication back to the excavator that the facilities on site have been marked or a status update has been clearly communicated – which may indicate marking was not able to be completed for one reason or another. In effect, a site may not actually be ready for digging, even if all utilities have made a positive response, because some may indicate additional time is needed. This clear process avoids miscommunication or misunderstanding and can help prevent a project proceeding prematurely. Because positive response is already in law in many places, it is something that can either be refined or reformed as new research comes to light.

New technological applications, including electronic positive response and enhanced positive response, can further improve communication, reduce misunderstanding, and ensure both damage prevention and excavation safety. Many states have updated their laws to specify these technological evolutions to standard positive response and many more may consider them. Further reform may center on response codes that can improve engagement. Positive response can be passive for the excavator, as they await hearing back from every utility and may often proceed with digging as soon as the last one confirms their work is complete. Many states require the excavator to verify the positive response, which is somewhat more active. In Virginia, response code 60 is used when an alternate locating schedule is needed.[26] In that case, the excavator has to proactively accept a proposed change, ensuring they are on the same page as the locator and all parties agree to amend the timing for excavation readiness.

Some states are also experimenting with enhanced positive response, where electronic platforms allow excavators to see not just a status code but photographs, digital manifests, or even facility maps, raising policy questions about what level of detail should be mandated or incentivized.[27]

On the other side, Locators can benefit from clearer instructions from excavators, exemplified by white-lining laws or practices, which can also be accomplished via technology through electronic white lining. Some Facility Notification Centers now support electronic white-lining, which can improve precision and reduce overnotification, but raises policy tradeoffs about revenue models, ticket scope, and fairness between excavators who can stay remote and locators still required to walk the site.[28]

Both enhanced positive response and electronic white-lining illustrate how evolving technology forces policymakers to revisit existing statutory frameworks – determining whether to codify new practices, incentivize them through grants, or simply allow them to develop as industry norms.

These kinds of refinements illustrate how the U.S. system is not short on rules or requirements; if anything, developments like enhanced positive response and electronic white-lining show how many layers of law and regulation already shape damage prevention in the United States. By contrast, many countries operate with only minimal statutory frameworks – or none at all – highlighting how unusual the U.S. landscape of 50 separate state dig laws really is.

An Embarrassment of Riches

Through all this discussion of law, many stakeholders may take for granted that in many ways, the law has shaped industry dynamics and how the process works. Even while industry best practices and self-regulating can fill in gaps, the overall structure of the industry rests on the framework of what states require. It may be surprising to learn that such robust law is virtually absent in some places. For years, many in Canada have lamented a lack of comprehensive legislation, with Ontario being the only province to have such an approach.[29]

How Ontario took up that legislation is a common experience – and one we explain more in the following chapter. It took a tragedy to lead to change. In that case, it was 2003 and a contractor was operating a backhoe on a roadwork site. When the machine struck a natural gas pipeline, it resulted in an explosion that killed seven people on scene and in a mixed commercial and residential building at a strip mall and injured four others.[30] As Mike Sullivan explains, "It was the proverbial straw that pushed One-Call legislation in Ontario forward."

Because a contracted locator failed to identify the pipeline that was struck, lawmakers were quick to draft legislation that would more clearly dictate what processes must be followed to prevent this type of tragedy again. Despite many tragedies in other Canadian provinces over the years, no similar legislation has been drafted.[31]

Around the world, law often changes because of an incident. In 2004, an explosion in Belgium changed damage prevention across much of Western Europe. When a construction vehicle struck a gas line that killed 24 people and injured over 130, policymakers acted quickly. The worst tragedy of its kind in over 50 years, there were reforms from many sectors, including security and construction to ensure it was the last. New survey rules were dictated and pipeline safety standards were rewritten. Other nations reacted as well, with France updating its laws to strengthen construction standards in the years that followed.

For Canada, the slow pace of legislation has in some ways also been a benefit for those thinking creatively. We have already explained how being disenfranchised from 811 led Canadians to move more swiftly to "clicks" rather than "calls." This both benefited their damage rates and future proofed them from the eventual decline of phone-number-based marketing. Another benefit has been that lack of regulation – while delaying stakeholder use of the system – has also meant no locking in of less desirable or inflexible standards.

"In Alberta's pursuit of comprehensive damage prevention legislation, we came to realize that it could be operationally stifling." Mike Sullivan noted. "For instance, USP may not have been able to shift from Calls to Clicks, implement the s Annual Member Fee (AMF) revenue structure or the Alternative Locator Program (ALP)."

While delivering a presentation on the alternative locator program at the 2025 Global Excavation Summit in Dallas, Sullivan's perspective was validated. During the presentation, one commenter noted the ALP was beneficial and clearly reduces damages, but wouldn't be possible to implement in the United States due to current legislation.

While laws across all 50 states have different hurdles and opportunities for innovative programs, this underscores the fact that designing legislation is tricky, and language needs to be used that doesn't "lock" the legislation or regulations in time.

Changing Laws

We've covered in this chapter that laws often change only in response to major incidents and that they can be otherwise slow to change in the ordinary course of business. In addition to no federal damage prevention laws – or a standardized nationwide dig law – which many stakeholders adamantly desire and many strongly oppose, the number and variety of dig laws at the state level are significant. This leads to a question about the pace of change and the driving forces behind it.

In most cases, state 811 Centers are central figures in identifying needs or opportunities for improvement, but are often merely one voice out of many. Illinois is an exception: JULIE, the state's designated facility notification system, has a formal role in reviewing and recommending legislative or regulatory changes.[32] By contrast, call centers in most other states can only advise, with any statutory changes left to the legislature or a public utilities commission. This distinction helps explain why reforms can take years to materialize, even when call centers and industry stakeholders broadly agree on needed updates.

In Michigan, the dig law (PA 174) gives the 811 Center Board of Directors the ability to implement system changes without going through the lengthy framework of legislative change unless it is needed. In many cases, like electronic white-lining, the center has been able to accomplish that change directly through the system with its stakeholders. MISS DIG CEO Nick Bonstell explained that, "I believe this is an optimal framework for organization structure to keep up with the pace of change." This likely explains why MISS DIG currently has a 90 percent electronic ticket percentage (with fewer than 10 percent calls) with the majority of those being the professional excavator entering electronic white-lining directly into the system.

The gradual evolution of state dig laws underscores how difficult it is for policy to keep pace with available technology. While smartphones, GPS, and digital mapping have become nearly universal, statutes and regulations in most states still reflect a framework rooted in mid-20th century practices. The Aii 2024 *Damage Prevention Report Card* found that in the eight years since its first edition, only a handful of states have made substantive updates to incorporate proven tools like electronic white-lining or enhanced positive response, and nearly all continue to under-leverage technology that is already commonplace.[33] This policy inertia contrasts sharply with the rising costs and risks of excavation damage, reinforcing the central point that safety depends as much on legislative urgency as it does on technical innovation.

Against that backdrop, the U.S. damage prevention system can be seen as both mature and with room to grow – mature in the breadth of its laws and oversight, yet uncharted in how to reform from here and how stakeholders will adapt to and create new realities.

With a robust federal canopy, continuous educational resources, grant funding, and the ability to conduct enforcement, there is a cohesive national damage prevention. Nevertheless, this is not a standardized or unified dig law. Some stakeholders believe more federal standards should be in effect, while others view the existing federal law as more than adequate in favor of diverse state approaches. While those state laws can leave gaps, there are so many states that stakeholders can learn from what works and what doesn't across the country to inform any reforms. In some cases, the law creates some of the tension in need of resolving and thus needs change. In other cases, narrow new laws are needed.

Importantly, many (and perhaps most) challenges in the damage prevention space do not require new laws or regulations. In fact, the member-driven process through the Common Ground Alliance and various other trade associations helps formalize consensus best practices that, when implemented, avoid many of the costs and inefficiencies that new rules attempt to revolve.

As the industry strives forward, whether through aspects of self-regulation or simply exceeding regulatory standards, best practices must also be refined. It will not be enough to simply achieve basic consensus that a practice is desirable, but the practice itself must be both data-inspired and consensus-driven. When evaluation of these practices accounts for variables and control groups and are tested under diverse regulatory regimes, they will truly live up to the name of best practices. From there, it will take dedicated leadership within the industry to ensure they are adopted.

As we have repeated throughout the previous chapters, culture is the most important aspect of safety and no law can create culture. Culture is also the only thing that ensures regular implementation of best practices throughout an organization.

Individual stakeholders can employ proven technological best practices, but sometimes that requires 811 Centers to offer and support certain options. It may require state law changes to make that move or it may require PHMSA grants to increase capacity. If and when all these stakeholders are aligned, real change can happen faster.

Chapter 12 Resources

ESA/ACTS Now Town Halls

Notes

1 Research and Special Programs Administration, Department of Transportation. (1990, September 20). Grants for state pipeline safety programs; state adoption of one-call damage prevention program. Federal Register, 55(183), 38688–38692. https://www.phmsa.dot.gov/sites/phmsa.dot.gov/files/docs/standards-rulemaking/rulemakings/archived-rulemakings/59736/55-fr-38688.pdf

2 Research and Special Programs Administration, Office of Pipeline Safety, U.S. Department of Transportation. (1999, August). Common ground: Study of one-call systems and damage prevention best practices. https://primis.phmsa.dot.gov/comm/publications/CommonGroundStudy090499.pdf

3 Common Ground Alliance (CGA). (2026, February). CGA Best Practices Version 22.0; History of the Common Ground Alliance. https://bestpractices.commongroundalliance.com/1-Introduction/102-History-of-the-Common-Ground-Alliance

4 U.S. Congress. (2001, December 20). H.R. 3609, Pipeline Safety Improvement Act of 2002, 107th Congress (2001–2002). https://www.congress.gov/bill/107th-congress/house-bill/3609

5 Federal Communications Commission (FCC). (2005, March 10). FCC Designates 811 as the Nationwide Number to Protect Pipelines, Utilities from Excavation Damage. https://docs.fcc.gov/public/attachments/DOC-257293A1.pdf

6 Occupational Health & Safety (OHS). (2007, May 4). Government, industry officials launch national "call before you dig" number. https://ohsonline.com/articles/2007/05/government-industry-officials-launch-national-call-before-you-dig-number.aspx

7 Pipeline and Hazardous Materials Safety Administration (PHMSA). (2023, February 16). Pipeline Safety, Regulatory Certainty, and Job Creation Act of 2011. https://www.phmsa.dot.gov/legislative-mandates/pipeline-safety-act/pipeline-safety-regulatory-certainty-and-job-creation-act-2011

8 Pipeline and Hazardous Materials Safety Administration (PHMSA). (2015, July 23). Pipeline safety: Pipeline damage prevention programs. Federal Register, 80(141), 43836–43869. https://www.federalregister.gov/documents/2015/07/23/2015-17259/pipeline-safety-pipeline-damage-prevention-programs

9 U.S. Congress, House Committee on Transportation and Infrastructure, Subcommittee on Highways, Transit and Pipelines. (2006, March 16). Pipeline safety: Hearing before the Subcommittee on Highways, Transit and Pipelines of the Committee on Transportation and Infrastructure, House of Representatives, 109th Congress, 2nd session (109-57). https://www.congress.gov/109/chrg/CHRG-109hhrg28272/CHRG-109hhrg28272.pdf

10 Pipeline and Hazardous Materials Safety Administration (PHMSA). (2024b, August 5). State Programs Overview. https://www.phmsa.dot.gov/working-phmsa/state-programs/state-programs-overview

11 Pipeline and Hazardous Materials Safety Administration, Department of Transportation. (2026). Protection of underground pipelines from excavation activity, 49 C.F.R. pt. 196. Electronic Code of Federal Regulations. https://www.ecfr.gov/current/title-49/subtitle-B/chapter-I/subchapter-D/part-196

12 U.S. Department of Transportation. (2025, March 17). Pipeline and Hazardous Materials Safety Administration: Key grant programs. https://www.transportation.gov/rural/grant-toolkit/usdot-competitive-grants-by-agency/phmsa

13 Pipeline and Hazardous Materials Safety Administration (PHMSA). (2025, March 6). One call grants. https://www.phmsa.dot.gov/grants/pipeline/one-call-grants

14 Pipeline and Hazardous Materials Safety Administration (PHMSA). (2021). Nine Elements of Effective Damage Prevention Programs. https://primis.phmsa.dot.gov/comm/DamagePrevention9Elements.htm

15 National Transportation Safety Board (NTSB). (2025, October). Investigation Report. https://www.ntsb.gov/investigations/AccidentReports/Pages/Reports.aspx?mode=Pipeline

16 National Transportation Safety Board (NTSB). (2023, December 14). NTSB retires most wanted list of transportation safety improvements. https://www.ntsb.gov/news/press-releases/Pages/NR20231214.aspx

17 Prothro, J. (2017, October). PHMSA Damage Prevention Initiatives. EPA. https://www.epa.gov/sites/default/files/2017-11/documents/9.prothro_2017aiw.pdf

18 National Conference of State Legislatures. (2023, December 11). How states protect pipelines from excavation damage. https://www.ncsl.org/energy/how-states-protect-pipelines-from-excavation-damage

19 Mississippi Code of 1972, § 77-13-27. (2024). Enforcement of damage prevention. https://www.msdamageprevention.com/law/

20 City of Chicago. (2021). OUC existing facility protection (EFP) process. https://www.chicago.gov/city/en/depts/cdot/supp_info/ouc–exising_facilityprotectionefpprocess.html

21 Alaska 811. (2024, February 20). Locates. https://www.811ak.com/locates/

22 Common Ground Alliance. (2022). 2021 DIRT Report. https://dirt.commongroundalliance.com/2021-DIRT-Report/Executive-Summary

23 California Legislature. (2025). Cal. Gov. Code § 4216. California Legislative Information. https://california.public.law/codes/government_code_section_4216

24 Common Ground Alliance (CGA). (2023, September). 2022 DIRT Report; Late Locates: A Current and Emerging Crisis. https://dirt.commongroundalliance.com/2022-DIRT-Report/Late-Locates-A-Current-and-Emerging-Crisis

25 Crawford, B. Scott. (2025, August). "Sound and Fury, Signifying Nothing": Moving Excavation Readiness Beyond Rhetoric and Toward Truth Through a Call for Dialectic. Alliance for Innovation and Infrastructure. https://www.aii.org/wp-content/uploads/2025/08/Sound-and-Fury-Signifying-Nothing.pdf

26 Virginia 811. (2024, January). Changes to CODE 60: Common Questions & Answers. https://va811.com/wp-content/uploads/2024/01/CODE-60-QA.pdf
27 Dierker, B. & Rogers, O. (2024, September). 2024 Damage Prevention Report Card: Tracking innovation in state dig laws to promote communication and collaboration among stakeholders. Alliance for Innovation and Infrastructure. https://www.aii.org/wp-content/uploads/2024/10/2024-Damage-Prevention-Report-Card.pdf
28 MISSDIG811. (2021, December 12). OneCallAccess ticket entry best practices. https://www.missdig811.org/cm/dpl/downloads/content/7134/Ticket_Entry_Best_Practices_-OneCallAccess_12-12-21.pdf
29 Canadian Common Ground Alliance (CCGA). (2012). Damage prevention legislation elements required for Canada. https://scga.ca/files/CCGA-WhitePaper.pdf
30 Global News. (2011, December 17). Enbridge fined $700,000 in deadly 2003 gas explosion. https://globalnews.ca/news/190592/enbridge-fined-700000-in-deadly-2003-gas-explosion-2/
31 Wall, D. (2020, March 2). Private land locates remain major concern, OSWCA delegates warned. https://canada.constructconnect.com/dcn/news/associations/2020/03/private-land-locates-remain-major-concern-oswca-delegates-warned
32 Illinois General Assembly. (2025). 220 ILCS 50/11: Penalties; liability; fund. Illinois Compiled Statutes. https://www.ilga.gov/documents/legislation/ilcs/documents/022000500K11.htm
33 Dierker, B. & Rogers, O. (2024, September). 2024 Damage Prevention Report Card: Tracking innovation in state dig laws to promote communication and collaboration among stakeholders. Alliance for Innovation and Infrastructure. https://www.aii.org/wp-content/uploads/2024/10/2024-Damage-Prevention-Report-Card.pdf

Chapter 13

Human Impact

> The scenarios in previous chapters present fictional but common visuals from across the damage prevention ecosystem. To illustrate the stakes more clearly, this introduction is a completely true story. Only the names, locations, and ancillary details are changed out of respect for the individuals and communities impacted. And of course, no negative implications are intended for the names, locations, or description that may bear a resemblance to actual entities.

While Nashville, Tennessee is a thriving metropolitan city, it also hosts dozens of suburbs and smaller towns. One Monday afternoon, IntelComms notified Tennessee utility notification center about a fiber installation taking place later that week in a residential neighborhood near Antioch. IntelComms had engaged Cityside Construction, LLC as a subcontractor to lay the fiber in the area, but IntelComms retained the contractual obligation to provide notice to the 811 Center and to remain in close communication with Cityside while they engage in the work.

Weather throughout the week was clear and sunny, with highs in the 70s and a slight breeze. No major construction projects were underway and every utility was sent a notification by the call center to either screen or mark their facilities. By Thursday, all the marking was completed.

Local water, telecom, and electricity utilities sent in-house and contract locators to the street throughout the week, each marking the path of the facilities. Volunteers Power, a local natural gas subsidiary of AmeriPower Energy, sent a locator on Wednesday, who both painted and flagged sections of a roadway and residential neighborhood. The Volunteers Power locator submitted his report Wednesday afternoon and the state's positive response portal accessible at the call center website indicated every affected utility had responded. Unbeknownst to the locator for Volunteers Power, he had missed the location of the capped gas main section just to the side of the line's route.

In the late afternoon on Thursday, workers with Cityside Construction were drilling to install the fiber lines. One worker felt the drill strike something below the surface. What he couldn't see was a four-inch polyethylene natural gas pipeline operating at 30 PSI. He immediately backed off and powered down his equipment, and two other crew members on scene could see faint waves in the air and began smelling the mercaptan odor agent added to natural gas.

The crew supervisor on site immediately pulled his men back several yards and called 911, instructing the drill operator to call 811 concurrently. Within 15 minutes, the line break was reported and local fire personnel were on scene. Inside of 30 minutes, a Volunteers Power

DOI: 10.1201/9781003796619-15

employee had also arrived to stop the leak. He did so by squeezing the off the main to stop the flow of gas.

Out of caution, Volunteers Power and the local fire department helped evacuate a local business within feet of the leak, which was occurring more or less in the street next to their parking and the side of their building. Several feet in the other direction, a sidewalk connected to a residential neighborhood, where homes stood another 50 to 200 feet from the leak. Those homes and other nearby businesses were not evacuated.

Squeezing the main did not completely stop the flow of gas. Less than three hours later, the home closest to the leak was leveled in a natural gas explosion and fire. Two neighboring homes were devastated with blown out windows, collapsed roofs, and fire scorched siding.

The quiet neighborhood never saw it coming. Even with the fire department activity earlier in the afternoon, there was no sense of what was to come. Denise Albright, who lived across the street and loved sitting on her porch in the mornings, certainly never expected that she would soon lose an entire home in her neighborhood. She couldn't have imagined that the sweet 5-year-old boy she used to see running up and down his driveway would die in the fireball, or that his 10-year-old sister would have her own bruises and burns. Their father, likewise injured.

Life would not return to normal on this quiet street for years. Not only did it experience a tragic loss of life, but the neighborhood was reshaped with entire structures missing. Emergency personnel taped the street off with bright yellow caution and police tap, federal investigators descended on the community. Even many months later, lives would still be further impacted.

The state attorney general filed a lawsuit against Volunteers Power for its failure to identify the gas main and therefore lack of compliance with state law. Life insurance, home insurance, business licenses and more would take years and onerous loops to jump through to settle and resolve. The weight of guilt of the power company employees, locators, contractors, and emergency response personnel will carry on for lifetimes. The little girl losing a brother in such formative years will ripple for decades.

Many people insulated from the tragedy itself were nonetheless impacted by it. Roads were shut down, traffic rerouted. Businesses lost revenue, utility rates slowly began to creep upward. The planned fiber installation would be delayed and promised services rendered later and with higher price tags. These and other ripples continued for months and years after even this small-scale tragedy. The one in this story is still unfolding. The victims and families, the communities impacted continue to grieve and feel loss. The companies involved face ongoing investigations, which, even when participated in with good faith, can still destroy livelihoods and lead to permanent inability to operate.

- - - - - - - - - - - -

State laws on damage prevention are almost always rooted in disasters. No one thinks about the bomb that doesn't go off, so there is little or no focus and energy to change laws when things seem alright. But when *a bomb* does go off, legislators are quick to act. With the newscycle at their back (not to be cynical, but it is natural to capitalize on public sentiment), lawmakers often follow the public outcry and attention to finally move bills that they would otherwise not waste their political capital on, such as updates to buried infrastructure excavation laws. It is always a tragedy when a major preventable incident occurs, but the silver lining may often be that needed change finally happens.

Damage to natural gas pipelines can have devastating effects. This is a real image, but different from the story above. Bay Area California- E Lewelling Gas Explosion and Fire.

Image from Shutterstock.

The human impact is enormous and not always captured purely in casualty counts. As previous chapters have demonstrated, the rippling effects of a damage can be just as life-altering as the incident at the epicenter. They almost always impact livelihoods, even if indirectly.

As discussed in a previous chapter, it was six key pipeline incidents in 1993 and 1994, all caused by excavation damage, that raised the alarm nationally and put into motion the stakeholder engagement that became the modern damage prevention industry.[1] Those six incidents shed further light on the human impact and why it was so important to form a robust damage prevention community – as well as why it remains so important to continue to improve today to prevent the next big disaster.

Taken together, the early 1990s incidents (which were by no means the only excavation damage incidents in this decade or even across the two years NTSB quoted) led to direct costs and fatalities that rippled through communities. The indirect costs soared in the immediate aftermath, and investigations took years to complete.

Ten people lost their lives from these six incidents alone, while as many as 200 were injured. One incident in Edison, New Jersey was so massive that phone company officials estimated that over 200,000 calls to 911 overwhelmed the system, jamming the lines and preventing bystanders from reaching emergency services in the hour after the incident.[2] Other incidents tell the story of the human impact from different perspectives and across many different sectors. While many of these have already led to critical reforms, other underlying dynamics are still in place and the potential for future incidents continues.

Peering into the details of a few incidents helps illustrate how excavation damage affects communities and the nation.

Human Casualties: Olympic Pipeline

Perhaps one of the most often-cited examples of an excavation-related pipeline tragedy took place in 1999 in Bellingham, Washington. The shock of the incident primarily related to the enormous fireball that lit a creek ablaze. The pipeline was carrying gasoline, which can quickly spread through fumes in the air, making ignition sources from a far wider vicinity extremely dangerous. In this case, over an hour after the rupture, fire burned approximately 1.5 miles along the creek downstream of the rupture itself.[3]

This tragedy also stands out because of its victims.[4] Certainly, many other excavation incidents have higher total fatalities and injuries, but in the case of the Olympic Pipeline rupture, all three were 18 years old or under.

Their deaths were also particularly tragic because of how they occurred. The first victim, an 18-year-old, died by drowning. He was in a canyon in the local creek fishing and the fumes knocked him unconscious, leaving him face down in the water. Every death is a tragedy and all excavation incidents are avoidable, but these circumstances seem to stack avoidable factors in layers that make the young man's death hit particularly deeply. Steep canyon walls in the creek made a quick exit nearly impossible in the face of rapidly advancing fumes. He likely would not have been able to make it out from the time he first smelled them until they overtook him. Had he made it out, surrounded by the fumes, he would have likely succumbed to fire. In the end, the excavation incident years earlier, rupture that morning, the environmental and even shape and depth of the creek all conspired to the young man's drowning, even though he simply left home that morning for recreational fishing, not knowing the chain of events put in place years earlier would take his life. All of these factors together are what makes families ask "why." Why the initial incident, why the release volume and radius, why their son?

The two other victims were both 10 years old. They were likewise playing near the creek. And the nature of their tragedy also has a high degree of emotional weight. Each boy survived the fire itself, but received severe and painful burns covering 80 to 90 percent of their total body surface. On-scene emergency responders ensured they were found and treated promptly, but the severe second and third degree burns to their heads, bodies, arms, and legs were catastrophic. They were quickly transported by helicopter to a burn unit in Seattle. It took nine to 14 hours for each boy to succumb to their injuries. The trauma and confusion of victims and bystanders alike must have been palpable.

Many incidents occurred prior to this and many more have since, but in 1999, this one demanded nationwide attention and congressional hearings.[5] Everyone could picture their children as the ones who passed away; the image of a huge smoke plume rising above a community made everyone feel their own vulnerability; and the hidden nature of so much gas and hazardous material infrastructure prompted a feeling of insecurity that such incidents could happen anywhere at any time.

This incident also helped illustrate the scale of harm that can ripple out from a single incident. Deaths and injuries are the most vivid and emotionally significant losses – and in this case eight additional people sustained injuries. Physical destruction is next, and impacts far more people. It was estimated that over $45 million in property damage arose from this incident, which includes the pipeline, the city water treatment facility, and the complete destruction of a residential home.[6] Incidents like this can mean family homes gone in an instant, businesses and inventory obliterated, the only reliable car a person has being rendered inoperable. These costs affect livelihoods and take time to recover from. Even if and where insurance is able to make someone financially whole, the process is anything but simple and efficient, and very often involves investigations, claims processing, and more that in effect add on to the costs rather than compensate for them.

The city's water treatment facility, while still standing following the eruption, was completely inoperable due to damage. This resulted in safe water concerns and additional manpower needed to manually treat water for the community. The nearby creek was flooded with over 237,000 gallons of gasoline as well, causing environmental harm to wildlife and the downstream community.[7] The visual aftermath was striking. Bright green trees packed alongside charred black trees tracing the shape of the creek. Like a scar only visible from the sky.

The psychological and mental effects of an incident like this also linger in a community for decades. In Bellingham, an elementary school now bears the name of one of the young victims, permanently enshrining his memory and this incident in the public record for all time. The families who lost children, the children who lost friends and classmates, and the long and slow recovery period are costs that are virtually impossible to calculate and far from what monetary values can encapsulate.

Once again, this and similar tragedies are so hard felt because they are avoidable. The Olympic Pipeline incident was not a single failure, but a layered problem that began many years before the eventual incident.

In fact, back in 1994, an excavation crew had struck the 16-inch steel pipeline while installing modifications to the city's water treatment plant.[8] The unreported and unaddressed incident led to a latent defect that contributed to the incident five years later. Despite computerized systems checking for pressure and leaks, alarms going off, and the pipeline being shut down and restarted, gasoline continued to flow for nearly an hour and a half before the pipeline rupture ignited. As hundreds of thousands of gallons of gasoline left the pipeline, passersby began to smell "incredible odor" and call 911.

Multiple residents reported petroleum odor and discoloration in the creek. Locals struggled to breathe and one person's dog was convulsing from the gasoline fumes permeating the atmosphere. Evacuation began nearby even before the ignition occurred – likely saving many lives. On scene emergency response professionals and fire captain upgraded the incident to "Hazmat 1" or a potentially controllable incident involving hazardous materials. Within minutes, the status was upgraded once more to "Hazmat 2" denoting a greater hazard potential.

Many emergency response personnel were already on site by the time the gasoline was ignited. This is very often not the case. In fact, this is a virtually unique excavation incident because the original root cause occurred five years earlier and from the time of rupture, nearly 90 minutes elapsed before a fire broke out. This allowed both for the extent of the damage – as more fuel was released and covered a larger area – but also allowed for emergency response, evacuations, and control of the scene ahead of the tragedy itself. Many factors pushed this toward being far worse, while many helped mitigate.

Due to the volume of gasoline in the creek, fires burned for over an hour, including brush along the creek path. Fully 25 acres were burned throughout the incident, spanning the creek, the water treatment facility, and nearby residences.

This incident was years in the making. That underscores a critical fact that bears repeating, even if you've read it elsewhere in this book: every nick, strike, or damage must be reported to the facility owner. The smallest ding to a pipeline can lead to a leak years down the road and a potential disaster.

The nation's eyes were on the Bellingham community for weeks. Many casualties were claimed by the incident, and even the environment bore scars due to the blast zone. But not all casualties are so visible. Some take only a single victim. Some take place in silence.

Silent Impacts: Hospital Oxygen Disruption

It was partly cloudy with a high of 45 degrees in West Haven, Connecticut early in 2022. An 88-year-old veteran was quietly resting in a medical center. A U.S. Marine, he had served honorably in the Korean Conflict and, once home, spent over 40 years as a steelworker. While his health was generally somewhat poor due to age and several other ailments, he was alert and in stable condition while on supplemental oxygen that January day. Near the Veterans Affairs medical facility, a construction crew was doing planned work. During the course of their excavation, a piece of equipment accidentally severed an oxygen piping system.[9]

Inside, patients utilized wall oxygen and portable oxygen equipment. Three days passed before the oxygen disruption was fully remedied. For the 88-year-old patient, the silent change was more than a disruption. He suffered oxygen deprivation that materially changed his medical condition. He was transferred to intensive care, but while portable oxygen was being provided and adjusted amid a delayed and unclear response, the Marine became unresponsive. Within only a few hours, he passed away.[10] For his daughter, two grandchildren, and four great-grandchildren, this meant no more visits or goodbyes. A simple nick of a pipe changed lives forever.

Many hospitals and medical centers utilize oxygen piping systems and related piped gas, as well as power, fiber, and more. These types of incidents are a reminder that it does not take an explosion to result in casualties. In fact, there are often indirect consequences that extend far wider than the site of an incident.

Digital Consequences: Telecom Outage

The team was assembled in a Sensitive Compartmented Information Facility (SCIF). From this secure room, the small group from the Department of Homeland Security was isolated from the world. Inside, they were discussing and deliberating over highly classified information. No one outside this room could peer into it, and anyone inside the room was unaware of what may be happening.

What was, in fact happening, on the streets of Washington, D.C. in the mid-2000s was road construction.

With a report due to Congress by the end of the day, it was crunch time for this team of security experts. Their focus was literally a matter of national security, and both time and secure internet connection were paramount. In the minds of this team, that wouldn't be a problem. They were in an impenetrable fortress and had their ducks in a row to meet their deadline. But the road crew outside had other ideas.

Within a moment, the fiber optic line providing the secure high-speed and connection was severed. Impossible to repair in time for the deadline, the team was forced to turn to alternative communications methods. Far more used to employing the latest technology, on this day, the team employed a secure fax machine to send their final report to Congress.

While not a total loss, fiber optic cables providing data and intel to other agencies can be far more consequential. While purely presented anecdotally here, we have heard stories and rumor that a severed fiber optic cable serving a CIA, NSA, or other intelligence agency could cost upwards of a million dollars a minute in downtime – not the financial costs of repair, but the indirect and opportunity costs that are nearly impossible to actually quantify.[11] But it isn't just intel agencies that rely on fiber, with similarly hard to quantify costs, the public can very often experience direct impacts from cut cables.

On a hot summer day in central Florida, a construction company was working around a utility easement. Possibly working on existing fiber or adding new fiber, a mishap resulted in severing a fiber optic cable. This resulted in over a million Floridians in two major counties losing their ability to call 911 for over eight hours.[12] In 2021, over 15 million 911 calls were placed in the Sunshine State, which works itself out to be an average of over 41,000 per day. For a million Floridians to lose access for almost a full day certainly meant delayed care, unattended emergencies, and possibly life-threatening dangerous situations. The recency of this incident likely means the full impact is unknown and not yet investigated.

There are so many reasons a person may call 911, and even a short outage has real human impacts. Check in on an elderly neighbor, domestic abuse, a house fire, a heart attack on the street, a theft or mugging, vehicle accidents, and more.

As our modern economy pivots more to digital reliance, including ecommerce, data processing, and even the internet of things connecting infrastructure, disruptions to fiber optic lines are having an outsized impact. Recent incidents have grounded hundreds of flights due to fiber optic lines being cut, and while too recent to know the full impact, some basic math reveals how two simple fiber lines being severed can run up nearly $80 million in society costs.

Grounded Because of the Underground

Everyone in the damage prevention industry is acutely in tune with the consequences of excavation work. It is an ongoing joke whenever something goes wrong to blame an excavation incident – even if jokingly. In fact, an internet connection went out during an interview for this book and we joked that it must have been a cut fiber line! That said, we always hate to be proven right. That's just what happened to Mike Sullivan, whom you've heard from throughout this book.

Just after attending the 2025 Global Damage Prevention Summit in Dallas, Texas, Mike posted a USAToday article about a radar outage delaying his flight.

"My trip home from the ACTS Now Summit has been delayed. Wouldn't it be ironic if the radar outage was caused by an uncontrolled excavation? #ClickBeforeYouDig" his post read.[13]

It wasn't long before he was validated, and was quick to post again sharing more details.[14] In a third post – which we emphasize here as social media posts, because it is a great way to continuously conduct awareness and education – Mike used ChatGPT to help estimate the overall impact.[15] AI and fiber are intimately dependent upon one another, they are also driving new infrastructure buildouts and leading to more excavation damage, making them a fascinating pair now and going forward for the industry to account for. We reproduced his estimates here with additional context and validation.

In September 2025, on a peak travel day of Friday, a contractor in North Texas severed two fiber optic cables. In close proximity to two critical airports – Dallas-Fort Worth's DFW and Dallas's Love Field – this resulted in system failures and technology outages the Federal Aviation Administration (FAA) was not equipped for. As a result, over 1,800 flights were impacted across two days. Those included at least 1,200 delays and over 600 cancellations.[16]

Anyone who has flown more than once or twice knows how frustrating a delay or cancellation can be. Putting a dollar value on frustration, however, is difficult, but other data exists to capture these impacts.

First, consider how many people were likely impacted by these issues. Using all conservative figures, we can take 130 passengers per flight to be an average, accounting for the international and smaller regional aircraft in daily use at these airports. The 1,800 total flights then represent at least 234,000 individual passengers. I assume at least one or two of these are mothers or fathers, sisters or brothers, or business people traveling to see people. We can't know how many others are truly impacted, so we'll stick with the 234,000 passengers alone.

Of those, the rough breakdown of delays versus cancellations would be 156,000 and 78,000. The FAA uses a recommended value of $47 per hour to assess traveler savings or loss due to travel times.[17] We can use this to apply an assumed two-hour loss to delayed passengers and six-hour loss to passengers facing cancellations. Suddenly, the costs start to come into focus.

The 156,000 passengers with an average two-hour overall travel delay represent $14.7 million in loss. The 78,000 canceled travelers facing six hours of average delay represent $22 million.

For passenger travel time lost alone, this represents a soaring *$36.7 million*. Yet at least two major categories of impact remain: airline operations and economic spillovers.

For the airlines, both direct and indirect costs accrue every time a flight is off schedule. Some of those direct costs include unused aircraft fuel, ground crew and aircraft repositioning, crew compensation and more. Indirect costs flow out from these naturally, including schedule disruptions, passenger and crew compensation defined by both airline policy and government rules, regulatory penalties and fines, and issues with crew hours of service. Reputational harm can also impact future business as a result, even when the underlying delay or cancellation was outside of the airline's control.

The FAA and other sources put these direct and indirect costs as high as $30,000 for canceled flights. Using a conservative $20,000 for cancellations and $3,000 for delays, those add up quickly.

The 1,200 delayed flights total to $3.6 million in estimated losses, while the over 600 cancellations would toll over $12 million. This puts airline losses *over $15.6 million* from just a day and a half disruption.

As we mentioned earlier, those nearly quarter million passengers are not all solitary people. They have lives, families, businesses, and more. To adjust the math to capture just a slice of what these disruptions would likely entail, we'll use a conservative figure of $100 per passenger. Some may have been traveling for sentimental and emotional reasons like a wedding or funeral, which is difficult to capture financially. Others may have been traveling to close multimillion dollar deals. Considering that just an Uber, airport meal, and basic incidentals likely reach near the $100 figure on their own, we can be confident this is a conservative estimate and not overly redundant with the estimate above.

At $100 per passenger, *$23.4 million* emerges as missed connections, business impacts, family disruptions, and overall economic spillovers.

Still more can be added, including environmental costs from fuel consumption, aircraft idling, emissions impacts, and more. Lost baggage, customer service complaints, and more likely influence many of these costs – all because of two simple fiber optic cables cut by a contractor miles from the runway. In total, from passenger time losses, airline operations, rippling economic costs, and more, *over $75 million* in waste, loss, and avoidable costs hit the DFW area in less than two days.

Good public services have multiple redundancies so that a single wire cut doesn't take down the entire aviation industry, but modernizing those systems is an ongoing task. Even with redundancies, sometimes the thing that is struck is still a core service provider. In those cases, the impacts can be far more national depending on exactly which lines get damaged.

And You Thought Your Grandma Hated the Internet

"Georgian woman cuts off web access to whole of Armenia."[18] While the most dramatic incident in total scale, this headline is illustrative of the nature of the impact. In this real-life story (all of the stories in this chapter being entirely true), a woman digging for copper wire ended up severing the fiber optic cable that served as the backbone to the internet for an entire country.

This incident took place well outside of the United States and our safe digging procedures, but the fact that it was a 75-year-old woman with a shovel, and not a construction crew, underscores that anyone, with any tool or equipment, can cause excavation damage to buried utilities. It also underscores the scale of potential impact. An individual rarely knows what is below their feet, and even when bright and fresh spray paint designates that service lines are there, the individual may have no idea of the downstream effects a damage may cause.

While one may sever a neighbor's home internet (as this author has done), one could end up severing the connection point to far more people, as the elderly Georgian *spade-hacker* did. In her case, 3.2 million Armenians were left without internet access for five hours.

Back in the United States, far more fiber optic cable is in the ground, and far more people are excavating every day than on the outskirts of the former Soviet nation of Georgia. That means more opportunity for a strike. Indeed, data from Common Ground Alliance (CGA)'s Damage Information Reporting Tool (DIRT) Report indicates that telecommunications infrastructure is the most-struck buried facility year after year.[19] While these strikes may not reach the scale of an entire country, repeated and regular strikes inevitably add up to major costs. Were it to happen, a single day of Amazon going off line could cost more than $1 billion – or put another way, one-thirtieth (1/30) the entire annual GDP of the country of Georgia.

Fiber optic cables form the backbone of the internet and downtime costs millions of dollars for very limited disruptions. One analysis shows that for a single minute of downtime, Amazon would lose nearly a million dollars, Google's Alphabet would lose over half a million, and Facebook's Meta would be out a quarter million dollars.[20] These corporate revenues may not inspire sympathy, but loss of 911 emergency response services, private or small business communications losses, government services, and more are all truly at risk as well.

As artificial intelligence and data centers take a more prominent role in society, more data is pulsing beneath our feet, heightening the scale of impact and risk from a strike. Could you afford to be the excavator on a slim margin just paying your bills on time, who caused millions of dollars in losses by an accidental telecom nick?

Increasingly, fiber optic cable is being laid underground to connect the modern world. As more fiber and other buried utilities are installed, it becomes more important to know exactly what's below the surface. That means accurate records, precise maps, and good coordination among those who own and operate infrastructure. Without that, even well-intentioned projects can set the stage for failure. The next case makes that clear.

When Poor Records Kill: The Importance of Mapping

In November 1996, the Humberto Vidal building in San Juan, Puerto Rico was destroyed in a catastrophic propane explosion, killing 33 people and injuring dozens more.[21] It was one of the deadliest gas explosions in recent U.S. history. The NTSB investigation revealed that the explosion was caused by a propane gas leak from an underground pipeline that had been severely bent by an installation of a water main. Critically, the construction crew responsible for the water main had received information on the location of the propane pipeline, but it was inaccurate. Believing the area to be clear, the water main was installed too close to the propane pipeline, which was bent and put under constant stress.

Several other failures contributed to the explosion, but the root cause was determined to be poor mapping by utility operators and poor coordination between parties.[22] Authorities had contacted the gas company for details of any underground utilities in the area, but the maps provided were inadequate and incomplete. Despite this, no one from the gas company ever visited the site of the water main during construction. Had proper mapping systems and coordination protocols been in place, it is likely that the gas explosion would never have taken place. Additionally, abandoned pipelines and other underground infrastructure were not sealed properly, facilitating the spread of the gas underground and into the basement of the Humberto Vidal building.

This explosion serves as a tragic reminder of the stakes and consequences from underground utilities. More than 30 people lost their lives in an instant because of disorganization, not to mention the dozens more who continue to suffer from the injuries and memories from that day. Although this incident took place after the 1994 NTSB workshop, it was another catalyst toward meaningful damage prevention reform.

While the Humberto Vidal explosion centered around propane and water main infrastructure, the underlying failure – poor records, bad mapping, and lack of site coordination – echoes in the fiber deployment world today. The more we build beneath our feet, the more critical it becomes to ensure that every utility, whether carrying gas or data, is precisely mapped and actively coordinated. That is especially true because a leak or incident – whatever the cause – can become a tragedy due to external circumstances.

The Ignition at Mill Woods

The leak was not his fault. In fact, Peter Clark was merely driving a truck the day that changed his life, and arguably a nation, forever. His truck did not hit a pipeline, he did not leave a construction site early after cutting corners. He simply drove a delivery truck down the road and unsuspectingly into a colorless cloud of propane.

Fatefully, what was up to that point merely a pervasive vapor leak, suddenly became a conflagration when his one-ton glass truck served as the ignition source that set the atmosphere ablaze.[23]

Peter Clark worked for a glass delivery company bringing windows to new home constructions, and on this sunny Friday morning, the 22-year-old hoped to get off at a decent hour. Pushing through his schedule instead of taking a potentially early lunch stop, he drove through the neighborhood of Mill Woods on the outskirts of Edmonton. At this time, in March of 1979, it was very much still *way out there*, and construction was actively building it into the residential and commercial developments seen today. In this relatively remote and open area, Peter made his way down the road and suddenly sensed something was very wrong. As he recalls, he "probably saw the vapor but thought it was smoke" because many in the area set small brush fires to pre-thaw the cold Canadian ground for construction work.[24]

After taking in a "very strong smell of propane," the truck stalled, and instantaneously "the world just went up around me in flames." Were it not for the surge of adrenaline that accompanied the inhalation of propane, he likely would not have so quickly escaped the vehicle and thrown himself into the roadside snow head first. He knew he had driven into the cloud and that backwards up the road was the only safe direction not to go further into the flames.

For Peter, the story extends not only to the emergency drive by a life-saving apprentice to the hospital, nor the nine months of hospitalization, but for a lifetime. Still a vocal advocate and reformer today, Peter happily shares with all who will listen the details of the day and ways to continue to improve safety.

For the community, this day went on to impact not only the dozens near the incident, nor hundreds in the small neighborhood adjacent to the fire, nor even thousands, but tens of thousands. In fact, nearly 20,000 residents faced evacuation and gut-wrenching uncertainty in the days ahead.[25] Once the initial fire was contained and remaining propane left to burn out, the extent of the leak had to be attended to. As the cloud that engulfed Peter made clear, the propane vapors were occupying any and every place they could find – which may have included sewage systems, pipes, and potentially homes.

Concerned that at any moment, a manhole cover, a home toilet, or even an entire building could go up in flames – as building pressure from the gas had already forced some of these open – the fire department initiated the largest residential evacuation to date in Alberta, which excluding wildfire incidents in recent years remains one of the largest population evacuations in the province's history.[26] Firemen were focused on preventing further ignition, making those small brush fires to thaw the winter-hardened soil a key priority. Some spent hours into the night extinguishing them to ensure no ignition sources were present as the propane cleared the atmosphere, and the sewage systems and pipelines could be checked for gas pockets or flushed out and the gas dissipated.

In the end, no additional flames took over, and no deaths were reported. Peter Clark bore the worst of it, experiencing third degree burns to over 30 percent of his body. It all started when a pipeline was struck by a contractor. As is often the case, an initial strike is not what leads to an immediate incident, but a weakened pipeline left to corrode. In this case, three distinct marks across the pipeline exterior made two weeks before the fire invited the corrosion that led to the leak of over 228,000 gallons of propane.[27]

Heavier than air, the propane gas leaked and pooled around the pipeline and into the open area. By the time Peter Clark drove through, it had amassed into an enormous and rising cloud. While the rest is history, one counterfactual raises the stakes even further: later that morning, a school bus filled with kindergarten children was scheduled for the same route.

Water Damage Isn't Insignificant

The entire damage prevention industry was created in response almost exclusively to pipeline incidents. From NTSB investigations to PHMSA jurisdiction, from media headlines to casualty counts, it is pipelines that rule the day. And not just "pipe" but hazardous liquid and gas pipelines. These are undeniably the most dangerous types of underground infrastructure to suddenly strike with heavy equipment. But others, as some anecdotes above have illustrated, can be quite consequential.

> … a cable looks like a cable … Electric and Gas are the two things that'll kill you instantly.
>
> (Kurt Youngs, Youngs Excavating, *ESA Town Hall*)

Indeed, both gas and electric lines can have the most direct risk of fatality and injury. Even while others like telecom lines are struck more often, it is often merely a home internet connection that is severed. Still, these can go on to affect thousands or more. One facility type is all but disregarded by the public as a safety risk or merely an afterthought, but it is one that is virtually always buried and connects to every home and business: water lines.

People often don't assume water line damage is consequential, but it can be. Two incidents paint the picture.

In Branson, Missouri in August 2025, a summer construction project was nearing its 22nd week when something went awry. Workers at the historic downtown construction site hit a high-pressure water main.[28] Two men were taken immediately to the hospital by ambulance. Days later, one succumbed to his injuries and passed away.[29] The project will experience weeks of delay, not only to repair the damage, but to pump out and prepare the site again for continued work.

While the incident is too recent for a completed investigation, witness testimony, news coverage, and images from the site reveal thousands of gallons of water flooding onto city streets. The city did not have an immediate estimation on when the main would be repaired and water service fully restored.

Several neighborhoods and businesses were impacted with water outages, while thousands of residents in the city experienced disruption of water services and were placed under a precautionary boil advisory, commonly issued when loss of pressure in a water line potentially leads to contamination intrusions.[30] Boil advisories do not only mean an inconvenience, it can be prohibitive to low income, elderly, or disabled people. Many daily activities require safe water that we take for granted in the United States. Intentionally setting water aside to boil for such routine tasks as cooking, drinking, and brushing teeth, is a hassle, and recommendations to even throw out ice made during the water outage, underscore how much attention must be paid by innocent community members when someone else made a mistake on a construction site. With multiple city areas included, even a hospital was in the boil advisory zone.

The details in Branson are still in the fog of recency. An incident from 2013 can provide more clarity.

In Minneapolis on a cold January day in 2013, a construction crew was hard at work. Paint was already on the ground, and a private contractor was operating a craw excavator. Within hours, the streets would be flooded and the crawler would be overturned and laid haphazardly on its side. The story of this water main break is not so much one of death and injury, but expansive public disruption.

Like gas clouds explained in previous sections, water searches for voids to fill and spreads until it hits an obstacle. After the contractor's backhoe struck a 36-inch water main that Thursday morning, an estimated 14 million gallons of water did just that.[31] It sought out every possible location to occupy, displacing residents, businesses, and halting city life for days.

Across a four-city-block radius, water flooded the streets. While much of the water made its way into the Mississippi River nearby, its pathway included federal buildings, lower parking levels, and businesses.[32] One worker reported seeing 10 to 12 feet of water flooding a parking deck.[33] Cars with headlights on – which desperate people quickly abandoned – floated through the water. In a "water, water everywhere, nor any drop to drink" example, the local businesses lost their access to running water due to the main's sudden outage. The major outage also impacted a far broader portion of the city as pressure dropped across the system.

The impacts in the immediate vicinity were obvious, as standing water occupied roadways for blocks. Those businesses, which also lost their water access, could not be operated due to flooding. The lost revenue and downtime for local companies included at least nine restaurants, a theater, a museum, and the post office. Multiple cars in the lower parking ramp were submerged and inoperable. All local government offices in the area also closed early.

Multiple downtown bus and transit routes were affected, potentially making real-time rerouting decisions that many people in need could not track or react to, impairing their mobility. Fortunately, despite the local hospital losing water pressure, it had functional backup systems in place.

On another day in another city, perhaps the water in the roads left to simply dissipate would be a nonfactor. But in Minneapolis, with a daytime high temperature of only 22 degrees Fahrenheit that day, this was a hazard that would continue to plague the downtown area and make road conditions unsafe. With overnight lows hugging 11 degrees, the remaining water froze despite city efforts to scrape off water and ice as a cleanup and mitigation measure.[34]

One contractor's mistake shut down the project, cut off part of the city's water supply, damaged equipment, destroyed property, forced businesses to close, and led to days and weeks of clean up. And no explosions or electrocutions were involved – only water. A quick search will reveal this is not only common in other cities, but even the same cities experience it repeatedly, with another Minneapolis water main break taking place this very year and doing irreparable harm to entire business inventories that may force them to shutter their doors.[35]

We have underscored here in many different ways how the human impact of excavation damage truly soars to affect people entirely outside the construction sector. But it is worth a final reflection on just what the costs can be to those excavators themselves. It is imperative that everyone disturbing the ground knows the stakes – which must include how their actions may affect others – but at the end of the day, self-preservation is our most natural instinct and the thing that is most likely to sway a contractor or any other person to adhere to best practices and follow the rules.

Impact to the Stakeholders Themselves

The focus here is on contractors, because ultimately most or all of the damage comes from their actions – even if it was not their fault. But it still bears noting that many other stakeholders experience danger, risk, and tragedy. Locators are killed on the road, excavators suffer fatalities from hitting a line, and utility workers in the trenches can face devastating collapses. These underscore why following best practices is essential. Ultimately, best practices will help avoid the community impacts and nationwide attention from a massive incident, but it also helps prevent many of the job site incidents, casualties, and fatalities.

Outside of explosions, fires, and electrocutions – perhaps the most dramatic on-scene casualty events – many excavators become victims after breaking ground. In fact, trench collapses – the walls of an excavation site caving in on construction crews – are not uncommon and lead to serious injury and death. In fact, between 2011 and 2018, the Bureau of Labor Statistics reports that a total of 166 workers died from trenching incidents.[36] Best practices along with federal and state regulations and standard technology and equipment can prevent this, but those working fast or skirting these standards can easily find themselves buried alive or crushed to death. In 2022, OSHA reported the highest single-year death toll in 18 years, with 39 trench collapse deaths. Fortunately, the data indicates a 70 percent decrease in such fatalities since.[37]

These types of incidents range, with some taking place instantly and the victims simply gone as soon as the walls fall in, while others take dramatic and harrowing hours.

> There were two people that died in a trench collapse. And one of those individuals was able to speak through a pipe in the ground for a couple hours to the firefighters, and then they brought his wife out to the site to talk to her one last time before he died. And that just really resonated with a group of us.
>
> (Perry Silvey, Safety Manager, BT Construction Inc. *ESA Town Hall*)

Accidents don't just impact the people injured or killed, they reverberate through communities and change people. In an earlier chapter, a trench collapse changed Eric's life and ultimately his marriage. Others leave families broken in other ways and businesses unsalvageable. But sometimes they can also drive meaningful change. The incident shared by Perry Silvey led to the creation of the Trenching and Excavation Safety Taskforce (TEST). Every tragic incident reminds the damage prevention industry that what they are fighting for is not just lower costs and incident liability issues, its life and death.

While these tragedies have a profound impact on the damage prevention community, the fact that everyone is working toward reducing accidents can be uplifting. What better way to honor the memory of fallen colleagues than to push forward in making positive changes.

> I can work in an industry where I just make money or I can work in an industry that can make money and also really help people's lives, save people's lives, and I love that.
>
> (Duane Rodgers, CEO of PelicanCorp)

There are many true believers in the cause across the industry. We can attest to dozens of individual conversations with young upstarts and industry veterans, and many considering retirement, who all feel personally invested in seeing the industry advance and safety improvements secured. On more than one occasion, people have expressly rejected making any more money from the industry in favor of selfless investment, coordination, and outreach to continue helping former colleagues and stakeholder groups.

But the fact remains, they are so invested because they've seen the dark days. They've been on the ground for major disasters. They've lost friends or colleagues. They've seen companies bankrupted, families severed, and livelihoods lost.

Positive impacts motivate damage prevention professionals of every kind to strive for improvement, attend conferences, participate in discussions, and advocate for change. It takes fortitude and vigilance to begin each day knowing that a mistake may not simply result in disappointment or discipline, but could cost someone their life. While internal disagreements may arise, no one in the industry loses sight of the ultimate goal: building a safer future.

The incidents summarized in this chapter represent only a small portion of the broader reality of utility strikes. Still, they reflect the types of impacts that occur regularly – and how devastating they can be. From life and death to lost productivity, from explosions to silent disruptions, and from national outages to individual tragedies, excavation damage takes many forms. These can affect entire communities and broader regions or a single individual, but almost always carry more than financial costs. Reading an accident investigation report can carry real emotional weight, even for someone far removed. That weight only grows heavier for those directly involved. With all these consequences in mind, the next section lays out the more formal and quantifiable costs – spelled out in black and white.

Counting up the Costs

Many of the human impacts of excavation damage have been stated and repeated throughout this book. Every incident that occurs, some new consequence or impact may arise to add to the cumulative list of potential impacts. But the extent and scale of impacts has long been understood. Back in 1995, well before many of the most notable examples, and when the linear mileage of buried facilities was cited as a mere 20 million miles, stakeholders were already putting out authoritative lists of many expected risks. This one from the Alberta One-Call Corporation[38] makes the information already discussed in this chapter more concise:

CONSEQUENCES OF DAMAGE TO BURIED FACILITIES

The potential consequences of damage to buried facilities range from minor service disruptions to catastrophic conflagrations and include

- loss of life
- personal injury
- environmental contamination
- evacuation of residential areas
- explosion, fire, flood, or toxic gas escape
- disruption of essential services
- inconvenience to the public
- third-party property damage
- damage to construction equipment
- contractor down time and loss of production
- loss of product and revenue
- costs to rehabilitate injured workers
- costs to repair damaged facilities
- costs to rehabilitate environment
- costs to repair or replace construction equipment
- police, fire, and ambulance costs
- lawsuits
- medical costs
- legal costs
- administration costs
- increased WCB assessments
- increased insurance premiums
- reduced credibility with public
- reduction in contractor's ability to be competitive
- fines
- jail terms

Every facility owner will experience some of these consequences every time a facility is damaged as will every excavator who damages a facility.

There are of course many missing bullet points from known incidents and many potential bullet points we could add by simply using our imagination and applying what we know about the nature of excavation damage. But this brief list is enough to demonstrate how lives can be lost or unalterably changed all from a simple miscommunication, failure to follow the correct process, or lack of best practice being employed.

> … even if it's not our fault, if we hit something, we've already lost, it's going to cost us in some sort of way.
>
> (Fred LeSage, Senior Risk Engineer, AXA XL. *ESA Town Hall*)

The way impacts ripple to impact innocent third party bystanders is why proactive reform is so critical. We have noted that policy and best practices are often crystalized or refined in the aftermath of a major incident, but that does not need to be the roadmap for the future. Enough cumulative incidents have occurred across every state and across decades to know how to prevent every damage every time. Not only are the direct stakeholders counting on it, but their families, the wider communities, regional economies, and even the nation as a whole.

Policymakers have to have the public interest in mind. They have to ensure the public safety and economy are protected, and they have to prevent cascading effects that may harm others. Financial costs from excavation damage – merely a slim proportion of overall impacts – clearly demonstrate why policy has a role in protecting communities and regions. As we have detailed earlier in this book, costs come in many forms, including directly, indirect, and societal costs. As a refresher and summary on everything prior to this, consider the following list of costs and how they add to the human impact – both for the excavator who may have caused an incident (or other stakeholder whose root cause is identified) and for the communities surrounding the epicenter.

This list is not truly comprehensive, but representative of the major impacts that are borne as financial costs included in a final sum:

Items that may or may not be collected

- External collection costs/agency commissions
- Barricades/traffic control
- Permits (city/county/state/provincial) to install replacement cables/pipelines
- Legal fees and litigation costs
- Exposing the damage for repair
- Materials used in repair
- Restoration of the area
- Actual cost of internal labor
- Heavy equipment used
- Generator/power equipment
- Food, lodging, and travel expense
- Emergency mobilization (contractor/locator)

Time

- Damage investigation, on-site and follow-up
- Internal staff collection efforts
- Out of service complaints
- Insurance resolution discussions
- Overtime for unexpected increases in workload
- Employee time/travel for depositions/trial

Overlooked/difficult to track

- Lost customers
- Customer loss of use (refunds/credits)

- Resolution of customer complaints
- Engineering/reengineering due to damage
- Establishing outage bridge to coordinate services interruption
- Support staff (3–20) for outage bridge
- Workload delays
- Future failure points (damage may weaken system and lead to future failure unattributed to 3rd party)
- Damage data capture and submission (software and/or manual)
- Emergency on call ticket notifications
- Facility owner records updates
- Reporting requirements (FAA, 911, PHMSA)

Soft costs

- Loss of brand confidence
- Negative public feedback
- Difficulty maintaining customer relationships, especially large businesses, with inconsistent services

Societal costs

- Loss of 911/emergency services
- Loss of commerce
- Employee downtime
- Road closures/traffic delays
- Property damage

In many incidents, most or even all of these costs compound upon one another. A simple excavation incident can easily become life-altering for hundreds or thousands of people.

Reforms Small and Large, Within and Outside the Industry

Whether in financial terms, human lives, infrastructure reliability, or more, the impacts from excavation damage do not stop when a telecom line is repaired from the tip of a shovel or a gas pipeline is repaired from a slow leak caused by the bucket of an excavator. As the list above outlines, many issues take weeks, months, or years to resolve and things never go back to normal.

> We're not doctors and we're not nurses, but we are providing a solution or a service that no one can understand until someone dies.
>
> (Duane Rodgers, CEO of PelicanCorp.)

State policymakers should be proactive to shape and improve the damage prevention industry. While reactive policymaking may be the fastest, it depends on a tragedy that is avoidable. Getting ahead of the next incident is essential for minimizing the human impact of excavation damage. There are clear public safety issues, clear economic stewardship concerns, and many other good governance reasons for policymakers to intervene. Working with stakeholders and listening to the challenges they articulate is the best way to make meaningful but narrow reforms that can improve safety and avoid tragedy. And outside of policy, there remain many opportunities to both prevent and mitigate damage.

While public policy reforms are one vital aspect, changes to the rules and regulations within an industry are not the full picture. It takes more robust and holistic reforms to keep every person safe and protect underground infrastructure from excavation incidents. Two notable examples illustrate this point, one that highlights prevention, and one that highlights mitigation.

In Washington, an investigation into the Olympic Pipeline incident uncovered a series of errors, including the failure to notify officials of excavation damage five years before the accident. The criminal settlement against the pipeline owners set aside 4 million dollars for the creation of the Pipeline Safety Trust (PST).[39] In the years since, the Pipeline Safety Trust has played a critical role as an industry watchdog, safety advocate, and educator.[40]

The Mill Wood incident, by contrast, led to reforms at the local, regional, and even international levels.[41] [42] Peter Clark explained that it's often the small things that lead to major incidents or that impact the severity of an incident. That means small and simple fixes can also prevent or strongly mitigate such incidents. One meaningful advancement that came out of the 1979 incident at Mill Wood was the realization that clothing was responsible for severe burns and fatalities in the wake of fire-related incidents.[43]

At the time, it was common for workers across industries to wear their normal clothing or company-provided uniforms, which were almost universally a type of cotton or polyester.[44] Yet these materials are susceptible to burning and can melt into the skin or exacerbate injuries. On a normal day, these would simply be normal clothes, but during a disaster, *they are hazardous*. A revolution in fire-resistant clothing eventually took place, where company uniforms, delivery clothes, and other professional and industrial work clothes relied either on chemically treated fibers to prevent ignition, natural fibers like wool, or inherently flame resistant synthetic fibers like Modacrylic (many work shirts and pants) or Aramid (which is used in Kevlar).[45]

Other reforms were not quite as individualized as the choice of clothing material. Incident command procedures and emergency operation centers were developed across localities and even the country of Canada to help coordinate information between first responders, emergency management professionals, local government, and relevant utility operators.[46] Many local fire departments professionalized quickly, learning from this major incident and almost certainly preventing and mitigating future incidents we now do not know the names of because of their more streamlined emergency responses.

Still, the most pertinent reform to this book is the coordination of utility operators to prevent a pipeline strike like the one that made Mill Woods famous. In fact, this very incident, among other things, influenced the creation of Alberta One Call – the provincial Facility Notification Center, operating for over 40 years and currently known as Utility Safety Partners.[47]

To say silverlinings can come from tragedies is an understatement. The new procedures, organizations, and other reforms that came from now infamous incidents at Bellingham and Mill Woods and many others helped pave the way for the current systems that prevent untold numbers of tragedies daily. These small victories amid tragedy are worth celebrating. Even larger commitments like the recent finding by Ernst & Young (EY) that 78 percent of global companies and government agencies have plans to increase safety-related budgets to reduce workplace incidents.[48] Nevertheless, we realize it may be hard to relate to how significant a change in worker clothing or organizational budgets is to real-life safety. So we close out the anecdotes in this chapter with one final spotlight.

Reimagining What's Possible

Whether or not policy changes overnight, and regardless of how devastating incidents can be for individuals and entire communities, there remains light and hope that can be seen in many stories. Even after tragic incidents, when individuals survive, that is a blessing to be thankful for and celebrate. Some people dedicate their professions to preventing damage, while others dedicate their entire lives to it. One example is Cliff Meidl.

After experiencing his own incident, his life has pivoted toward selflessness, courage, conviction, and education.

Cliff couldn't tell you if he was alive or dead. The others on the construction site couldn't tell either. One moment, Cliff was operating a jackhammer and the next, he was laying on his back convulsing and with smoke wafting up from his clothing.

It was on a construction site in November of 1986. A plumber's apprentice, Cliff already had almost two years of experience under his belt coming into the day. Still, the 20-year-old had a lot to learn. While it was November, it was also Los Angeles, and the day was cool but perfect for the job he had ahead of him – breaking up concrete. With temperatures still rising from an overnight low of 55 degrees, the forecast called for all sun.

Cliff's job was operating a jackhammer. He was working in a small ditch in an excavation area on the site. For nearly 15 minutes, everything went according to plan as he chipped away at the concrete. Then, in an instant, the mechanical tip of his jackhammer struck three high-voltage electrical cables, each carrying 4,160 volts, sending nearly 30,000 volts of electricity through the equipment and straight into his body.[49]

Within a moment, he went into cardiac arrest. His limp body then slid back into the ditch, where the energized jackhammer was still in contact with the live wire. Virtually lifeless, his knees were pinned to the power tool with the voltage continuing to surge out from the electrical cables, giving a second round of high voltage assault on his body. As the electricity emanated through his body, it burst outward, creating exit wounds and grounding burns over nearly 15 percent of his body.

Toes on his right foot were blown off from the electrical blast, while his knees sustained significant damage from contact with the jackhammer and receiving the voltage directly. Nearly a third of each knee was burned away in the incident.

Were it not for the proximity of emergency responders with the fire department, Cliff's story would have ended there. But within minutes, emergency medics were on scene to revive him. Even after his arrival at the hospital, smoke was still rising from his smoldering knees, while electrical burns covered 15 percent of his body.

In all, nearly 15 reconstructive surgeries were needed for his legs and painful years of rehabilitation.[50]

Facing near-certain amputation, Cliff underwent a state-of-the-art reconstructive procedure performed by Dr. Malcolm Lesavoy, whose groundbreaking work ultimately saved his legs

While Cliff was the only casualty that day, the incident underscores how a single life can change in a moment. How many years it would take to bring him back the kind of standing, walking, and mobility so many take for granted. And while Cliff was the only one injured, the mental and emotional toll of those on the site and his family adds to the ripple. Of course, repair and project delays likely come with such an incident, but for just Cliff alone, untold thousands in costs accrued over the course of hospitalizations, surgeries, therapies, and rehabilitation. It may have been nearly five years before the final cost of all direct and indirect costs from his single incident was known. And while the costs were high – something even more significant came from it too.

Still in his early 20s, Cliff was regaining his mobility, still never able to run again, and fighting enormous mental hurdles each day to simply return to normal. But he would not let normal be his end goal. Instead, through incredible perseverance, Cliff began training and working in athletics for rehabilitation, ultimately pushing his body beyond what he may even have been able to achieve had no incident slowed him down or created such motivation to give him this drive.

Inspired throughout his recovery by Greg Barton – an American athlete only a few years older than Cliff who was still actively competing and winning Olympic gold medals despite a birth defect leaving him with two club feet – Cliff took up the sport of canoeing and kayaking.

Taking up the sport in his adulthood, Cliff became stronger and more skilled in the lead up to the 1996 Olympic games, winning a bronze, silver, and gold medal in the run-up U.S. Olympic Festival in 1995 hosted in Denver, Colorado. With this springboard, Cliff qualified for and competed in the 1996 Olympic Games in Atlanta, Georgia. He went on to compete in 2000 at the Sydney Olympics, where his fellow Olympians nominated him for the prestigious honor of bearing the U.S. flag during the opening ceremonies. Serving as a flag bearer is considered one of the highest honors an Olympian can receive

While he did not collect medals at these games, he had already won out – he survived, he regained his mobility, he had become an internationally competitive athlete, and competed in multiple Olympic Games! Following his 2000 Olympic Games, Cliff also received far more media attention than ever before, which further boosted his momentum into his next career path.

Recalling later, Cliff jokes that looking back on his long career in safety advocacy and motivational speaking that, "I got into safety *by accident.*" If not for the terrible event in 1986, Cliff may have gone on to a relatively standard life and no one would know his name, but many people also would not know about 811, the buried infrastructure below our feet, and the incredible danger that digging without following all proper laws and safety procedures can pose. Cliff's holistic approach to advocacy does start with basic awareness, but it includes the critical aspect that *construction safety* no matter where or on what job *is excavation safety and damage prevention.* Whether it's a private site or the work only involves minimal intrusion, all of it requires advanced notice to a Facility Notification Center so that every local utility is aware and can mark their lines.

Cliff represents far more than a silver lining. From clutching his life back from the jaws of death to waving the stars and stripes on behalf of a nation at the world's most competitive athletic stage, he is a testament to resilience. That he has chosen to use his story and the attention it brings to redirect attention to safety, awareness, and education is as critically needed as it is honorable.

Human Impacts – Human Solutions

The odds are that every person reading this book has been impacted by excavation damage. It's a virtual certainty in the United States given the expansive state of buried infrastructure, consistent construction activity, and known scale of annual incidents. Even those that do not make the headlines impact cities, states, and regions. Power outages, internet disruptions, traffic delays and reroutes, and more – even if you didn't realize it at the time, you have been party to one of the broader societal costs.

For many readers, this hits far closer to home – a friend, colleague, or family member who has been injured or witnessed an incident on the job site. Some may even have been injured themselves. And of course – hopefully far fewer – have experienced the worst loss.

Because everyone has been impacted, everyone has a stake in improving damage prevention – every person planning a DIY project; every person who works from home or plays video games to unwind; every person with water, gas, electric, or other services coming into their home, office, or business. Every person can spread this word, share the stakes, and even join the industry in a new career pivot or supporting role.

This industry needs passionate people. And it takes all kinds. One does not have to experience an incident to become a passionate advocate for improvement, reform, and education – but the more who have personal experiences, the stronger the education will be as well. As young people or those transitioning from other careers or industries consider damage prevention, the opportunity to advance safety is an endless frontier. Until there are zero damages, this vital profession will continue to be paramount – and even the day zero damages is realized, it will require the bulwark of this professional industry to maintain zero damages.

So what are you waiting for?

Chapter 13 Resources

ESA/ACTS Town Halls

Internet Outage Costs

Olympic Pipeline Tragedy Video

The True Cost of Water Damages

Trenching and Excavation Safety Taskforce (TEST)

Notes

1 National Transportation Safety Board (NTSB). (1997, December). Protecting public safety through excavation damage prevention (Safety Study No. NTSB/SS-97/01). https://www.ntsb.gov/safety/safety-studies/Documents/SS9701.pdf
2 National Transportation Safety Board (NTSB). (1995, January 18). Addendum (reconsideration request) to: Texas Eastern Transmission Corporation natural gas pipeline explosion and fire, Edison, New Jersey, March 23, 1994 (PAR-95-01). https://www.ntsb.gov/investigations/AccidentReports/Reports/PAR9501.pdf
3 National Transportation Safety Board (NTSB). (2002, October 8). Pipeline rupture and subsequent fire in Bellingham, Washington, June 10, 1999 (Pipeline Accident Report No. NTSB/PAR-02/02). https://www.ntsb.gov/investigations/AccidentReports/Reports/PAR0202.pdf
4 McClary, D. C. (2003, June 11). Olympic Pipe Line accident in Bellingham kills three youths on June 10, 1999. https://www.historylink.org/File/5468
5 U.S. Congress, Senate. (2000). Pipeline safety. Congressional Record, 146, 2189–2191. GovInfo. https://www.govinfo.gov/content/pkg/CRECB-2000-pt2/html/CRECB-2000-pt2-Pg2189-3.htm
6 National Transportation Safety Board (NTSB). (2002, October 8). Pipeline rupture and subsequent fire in Bellingham, Washington, June 10, 1999 (Pipeline Accident Report No. NTSB/PAR-02/02). https://www.ntsb.gov/investigations/AccidentReports/Reports/PAR0202.pdf
7 City of Bellingham. (2020). Olympic pipeline tragedy. https://cob.org/services/environment/restoration/olympic-pipeline-incident
8 National Transportation Safety Board (NTSB). (2002, October 8). Pipeline rupture and subsequent fire in Bellingham, Washington, June 10, 1999 (Pipeline Accident Report No. NTSB/PAR-02/02). https://www.ntsb.gov/investigations/AccidentReports/Reports/PAR0202.pdf
9 VA Office of Inspector General. (2023, July 27). Facility leaders' failures in communications, construction oversight, emergency preparedness, and response to an oxygen disruption at the West Haven VA Medical Center in Connecticut (Report No. 22-01696-160). https://www.vaoig.gov/sites/default/files/reports/2023-07/VAOIG-22-01696-160.pdf
10 VA Office of Inspector General. (2023, July 27). Facility leaders' failures in communications, construction oversight, emergency preparedness, and response to an oxygen disruption at the West Haven VA Medical Center in Connecticut (Report No. 22-01696-160). https://www.vaoig.gov/sites/default/files/reports/2023-07/VAOIG-22-01696-160.pdf
11 Gardner, A. (2009, May 31). Metro Dig at Tysons Stirs Underground Intrigue. https://www.washingtonpost.com/wp-dyn/content/article/2009/05/30/AR2009053002114_pf.html
12 McLeod, E. (2025, July 9). A cut cable caused a 911 outage in 3 Central Florida counties. how emergency calls kept coming. https://www.clickorlando.com/news/local/2025/07/09/a-cut-cable-caused-a-911-outage-in-3-central-florida-counties-how-emergency-calls-kept-coming/
13 Sullivan, M. (2025, November). My trip home from the ACTS Now Summit has been delayed. Wouldn't it be ironic if the radar outage was caused by an uncontrolled excavation?. LinkedIn. https://www.linkedin.com/posts/mike-sullivan-81a2a44_radar-outage-prompts-flight-delays-in-dallas-activity-7374985944304824320-tzpG?
14 Sullivan, M. (2025, November). Experts question FAA safeguards after cable cut causes Newark radar outage. LinkedIn. https://www.linkedin.com/posts/mike-sullivan-81a2a44_experts-question-faa-safeguards-after-cable-activity-7376298530656808960-etmV?
15 Sullivan, M. (2025, November). On Friday, Sept 18, 2025, a telecommunications cable providing radar services to Dallas Fort Worth and Love Field airports failed as a result of "two cut fibre optic cables." LinkedIn. https://www.linkedin.com/posts/mike-sullivan-81a2a44_on-friday-sept-18-2025-a-telecommunications-activity-7376822186881269761-N5pK?

16 Parker, J. (2025, September 23). Tech outage snarls hundreds of flights at Dallas-area airports. GovTech. https://www.govtech.com/transportation/tech-outage-snarls-hundreds-of-flights-at-dallas-area-airports
17 Federal Aviation Administration (FAA). (2024). Treatment of time. U.S. Department of Transportation. https://www.faa.gov/sites/faa.gov/files/regulations_policies/policy_guidance/benefit_cost/econ-value-section-1-tx-time.pdf
18 Parfitt, T. (2011, April 6). Georgian woman cuts off web access to whole of Armenia. https://www.theguardian.com/world/2011/apr/06/georgian-woman-cuts-web-access
19 Common Ground Alliance (CGA). (2025). 2024 DIRT report. https://dirt.commongroundalliance.com/Portals/9/Common-Ground-Alliance-DIRT-Report-2024.pdf
20 Baghel, S. (2025, August 1). How much do Microsoft, Apple, Google, meta earn per minute. https://www.bhaskarenglish.in/tech-science/news/how-much-do-microsoft-apple-google-meta-per-minute-your-favorite-tech-companies-make-more-money-than-most-people-see-in-a-year-135578823.html
21 National Transportation Safety Board (NTSB). (1997, December 23). San Juan Gas Company, Inc./Enron Corp. propane gas explosion in San Juan, Puerto Rico, on November 21, 1996 (Pipeline Accident Report No. NTSB/PAR-97/01). https://www.ntsb.gov/investigations/AccidentReports/Reports/PAR9701.pdf
22 National Transportation Safety Board (NTSB). (1997, December 23). San Juan Gas Company, Inc./Enron Corp. propane gas explosion in San Juan, Puerto Rico, on November 21, 1996 (Pipeline Accident Report No. NTSB/PAR-97/01). https://www.ntsb.gov/investigations/AccidentReports/Reports/PAR9701.pdf
23 Ivanov, J. (2019, March 3). Edmontonians mark anniversary of Mill Woods Explosion which resulted in safety improvements in Alberta. https://globalnews.ca/news/5016292/edmontonians-mark-anniversary-mill-woods-explosion/
24 Edmonton Journal. (2019, March 1). 40 years after the Mill Woods Explosion [Video]. YouTube. https://www.youtube.com/watch?v=2IfPoYgRHl0
25 Bell, D. (2023). Never to forget: Mill Woods' explosive history. https://yeg.tracksintime.ca/never-to-forget-mill-woods-explosive-history/
26 Bell, D. (2023). Never to forget: Mill Woods' explosive history. https://yeg.tracksintime.ca/never-to-forget-mill-woods-explosive-history/
27 Cole, C. C. (2015, July 15). The 1979 gas explosion: "The whole of Mill Woods could have blown up." Millwoods History. https://millwoodshistory.weebly.com/uploads/2/5/3/0/25302335/1979_gas_eplosion_july2015.pdf
28 McCaulley, S. (2025, August 21). Main Break shutdown streets, injures two in downtown Branson. https://www.branson4u.com/news/local-news/main-break-shutdown-streets-injures-two-in-downtown-branson/
29 KTLO. (2025, August 24). Worker dies after water main break in Branson. https://www.ktlo.com/2025/08/24/worker-dies-after-water-main-break-in-branson/
30 Serykh, M. (2025, August 21). Precautionary boil advisory. https://www.bransonmo.gov/DocumentCenter/View/21335/Precautionary-Boil-Advisory
31 Smith, K., & Smith, M. L. (2013, January 4). A deluge from below: Burst water main floods downtown MPLS. https://www.startribune.com/a-deluge-from-below-burst-water-main-floods-downtown-mpls/185599962
32 BMTN. (2018, March 8). Minneapolis water main break shuts down businesses, interrupts traffic - bring me the news. https://bringmethenews.com/minnesota-news/downtown-minneapolis-loses-some-water-service-after-main-break4
33 Williams, B. (2013, January 3). Water main break floods downtown Minneapolis | MPR News. https://www.mprnews.org/story/2013/01/03/water-main-break-floods-downtown-minneapolis
34 Smith, K., & Smith, M. L. (2013, January 4). A deluge from below: Burst water main floods downtown MPLS. https://www.startribune.com/a-deluge-from-below-burst-water-main-floods-downtown-mpls/185599962
35 Williams, B. (2013, January 3). Water main break floods downtown Minneapolis | MPR News. https://www.mprnews.org/story/2013/01/03/water-main-break-floods-downtown-minneapolis
36 Department of Labor. (2023, April 10). US Department of Labor, state agencies, industry leaders launch campaign to educate, alert Midwest employers, workers of deadly excavation hazards. https://www.dol.gov/newsroom/releases/OSHA/OSHA20230410

37 Occupational Safety and Health Administration (OSHA). (2024, November 4). Department of labor encouraged by decline in worker death investigations. https://www.osha.gov/news/newsreleases/osha-national-news-release/20241104
38 Utility Safety. (1995, January 25). DAMAGES TO BURIED FACILITIES. https://utilitysafety.ca/wp-content/uploads/2022/04/damages.pdf
39 United States Attorney's Office, Western District of Washington. (2003, June 19). Pipeline managers sentenced to jail in criminal case arising from the June 1999 gasoline pipeline rupture in Bellingham, Washington; pipeline companies ordered to pay $36 million in combined fines and penalties and to implement multi-million dollar safety plans. https://pstrust.org/docs/usdoj_doc1.pdf
40 Pipeline Safety Trust. (2020). About the trust. https://pstrust.org/about/
41 EAPUOC. (2023). The history of EAPUOC. https://eapuoc.com/news/the-history-of-eapuoc
42 UK Heath and Security Executive. (2019). Pipeline failure inquiry - mill woods area, Canada. March 1976. https://www.hse.gov.uk/comah/sragtech/casemillwoods76.htm
43 Smith, G. R., & Clark, P. (1992, January 1). The introduction of fire resistant workwear programs in Alberta. The Introduction of Fire Resistant Workwear Programs in Alberta. https://store.astm.org/stp19182s.html
44 Lovasic, S. L., & Neal, T. E. (2002). Industrial Flash Fire and Burn Injury Fundamentals with an Instrumented Manikin Demonstration of Protective Clothing Performance. https://nascoinc.com/content/resources/Industrial-Flash-Fire-and-Burn-Injury-Fundamentals_ASSE-.pdf
45 National Fire Protection Association. (2023). NFPA 2112 standard development. https://www.nfpa.org/codes-and-standards/nfpa-2112-standard-development/2112
46 Ivanov, J. (2019, March 3). Edmontonians mark anniversary of Mill Woods Explosion which resulted in safety improvements in Alberta. https://globalnews.ca/news/5016292/edmontonians-mark-anniversary-mill-woods-explosion/
47 Yahoo! News. (2019, March 2). Firefighter recalls explosion, fear that rocked mill woods in 1979. https://ca.news.yahoo.com/firefighter-recalls-explosion-fear-rocked-140000645.html
48 Tyson, J. (2025, September 8). Most firms aim to boost budgets for worker safety in next three years: EY. https://www.cfodive.com/news/firms-aim-boost-budget-worker-safety-three-years-ey-hr-esg-health-environment-OSHA/759548/
49 Deseret News. (2000, September 14). U.S. flag-bearer grateful for role, glad to be alive. https://www.deseret.com/2000/9/14/19780445/u-s-flag-bearer-grateful-for-role-glad-to-be-alive/
50 Nokes, D. (2013, Spring). The fire within: Cliff Meidl's campaign to improve workplace safety. Western Energy. https://georgia811.com/Media/WEI-CliffMeidl.pdf

Conclusion

Every day, nearly 350 million Americans go about their routine mostly ignorant to what is going on around them. It isn't their fault; everyone has his or her own life to manage – rent to pay, kids to wrangle, proposals to submit, errands to run. If everyone spent their days thinking about the world around them, nothing would get done, and everything would come to a halt. But through specialization, each person can worry about their own priorities while someone else takes care of things they take for granted.

John Smith doesn't have a green thumb and couldn't farm or garden to save his life, but he knows accounting and drives by a local grocery every day, where Kevin Asher stocks shelves, and Janet Ritmore manages the store. Violet Turner doesn't even know where electricity comes from, but she is the principal of a local elementary school that cares for and educates the whole city's children while the parents go to their day jobs. Everywhere we look, we see people who are experts at the thing they do and totally at a loss about many other things. Benjamin may not know how to fix his car, and his mechanic doesn't know how to prepare a legal filing – fortunately they can rely on each other.

Perhaps least considered above all is the specialization of damage prevention. If electricity, water, and gas do come to mind periodically for some people, the likelihood that they also picture men and women protecting the infrastructure itself is slim to none. Yet these professionals are ensuring reliable services and preventing catastrophic losses every single day.

These damage prevention professionals also have their own specializations – some conduct safe excavation, carefully breaking ground to build new infrastructure; some use hi-tech devices to locate buried facilities; others coordinate detailed communications between stakeholders; and others are even more specialized within their ecosystem. Importantly, they all act with a unified purpose – to prevent damage and ensure both worker and public safety.

It may be tempting to think some of these are just menial tasks or even *dirty jobs* showcased by Mike Rowe. Some may write them off as low skill or redundant with modern technology, but the truth is that the industry as a whole, and the individual roles within it, are bonafide professions. They have career trajectory and mobility, they offer rewarding and purposeful motivation, and they leverage and utilize cutting edge and evolving technology to unlock continuous improvement and opportunities. As more cross-disciplinary certifications arise, the professionalization of the damage prevention industry only crystallizes and strengthens.

As we have laid out through various stories, background, and quotes, this industry is also high-stakes. While a flat tire may ruin a day, and you can rely on a mechanic to patch it up, when

DOI: 10.1201/9781003796619-16

it comes to damage prevention, the modern infrastructure below the surface has much weightier consequences and broader impacts should something go awry.

The work these professionals carry out every day is literally the difference – in many cases – between life and death. It is also the difference between enormous economic consequences or smooth and economy-growing activity.

As the concluding chapter made clear, life is literally on the line every day. Quiet households and small businesses may be entirely unaware that they could be entirely wiped out of existence at any moment. Modern life is interconnected with power, water, gas, and other services we often depend on but take for granted. A normal morning routine of changing a thermostat, jumping in a shower, and using a stovetop to cook an egg means that outside and below the surface, thin plastic pipelines are sending colorless but highly flammable methane gas throughout family-friendly neighborhoods, parks, streets, and businesses. Virtually every square foot of these areas poses the same risk: an unplanned or poorly executed digging project can create a fireball.

Adding in the other buried utility lines and other facilities, and you get 50 million miles of underground infrastructure in the United States. That makes it clear that every dig, in every location, at every time has a risk. Yet the risks are not all equal.

Even for incidents that do not rise to the level of disastrous explosions, individuals or communities may still experience electrocution, contamination, flooding, and more. In other words, there is a range of impacts, which while perhaps not life threatening, may still be life-altering. Service disruptions can mean lost jobs, project delays, traffic rerouting, and inconvenience. Companies may have their reputation impacted so much they go out of business. Regulatory fines and litigation can bankrupt small businesses and tie construction sites up in a knot.

Finally, the economy – an aggregate of all these impacts – absorbs more than $150 billion every year from incidents involving striking buried facilities.

These consequences are all avoidable – and very often are avoided – when professional damage prevention workers ensure safe, smooth operations. Every day that passes without a house erupting in flames is a day that countless professionals – from project owners and utility companies to locators and contractors – successfully coordinated to build, protect, and maintain our infrastructure, all while ensuring everyone made it home safely. Every day, damage prevention professionals are literally holding back disaster. They stand between death and destruction, billions of dollars of losses, and life-altering infrastructure disruptions. And you can too!

As a homeowner or just average Joe, you can engage with the damage prevention process by simply knowing about and using 811. You can tell friends and family about the spray paint markings and what they mean when you come across them in the park, your street, or outside a business. You can ensure flags and stakes stay where they ought to be by asking children and others not to remove them without permission. And you can continue to learn about this industry through books, articles, and more – little by little helping it grow and ensuring more people know to click or call before they dig to protect lives and livelihoods.

If you want to join this dynamic industry, there are ample job opportunities and rewarding career paths. Not only can young people help shape how the damage prevention profession grows and evolves, but you can develop and deploy new technology and modernize how communication and collaboration take place. Regardless of your entry point into the industry – from digging small holes as a side gig to being a contract locator, from office internships to public works apprenticeships – every day, there is new mobility to rise up or make lateral moves within the ecosystem. With every move, you can learn new skills, broaden your perspective, and add value to the role you occupy.

> You start looking at our new generations ... they want to be part of something bigger than just a job ... damage prevention is so ripe for people that have got that kind of energy.
>
> (Debbie Shelly, CEO of Global Training Centre. *ESA Town Hall*)

Already in the industry? You too can make significant improvements. By helping damage prevention professional certification become reality, you can bring together your years of experience and draw in greater multidisciplinary learnings. You can attend more conferences, join local or national committees, and speak up at events and townhalls. Continuous education will make you better at your job and help those around you as you naturally share what you know. Joining organizations like Common Ground Alliance (CGA) or sharing your perspective with local chapters of other trade associations or nonprofits gives you more opportunity for networking, new career options, and the chance to identify younger professionals in need of mentoring whom you can help bring up in the industry. Training up the next generation of leaders is as noble a calling and vital need as any, and within damage prevention, not only does historical knowledge need to be passed down, but fresh thinking must always be infused. Only through multi-generational collaboration is this dynamic achievable.

> So we [have] got to continue to work together. And when we get through working together, we will be able to see the solution. We may not see it from where we are, but if we keep walking down the road, we'll see it just over the hill.
>
> (Roger Cox, founder of ACTS Now. *ACTS Town Hall*)

- - - - - - - - -

Scott's own story is one that shows a zig zagging pattern all throughout the industry. Since starting out, the ecosystem has only professionalized further, adopted more technology, and built up more cross-sectoral certifications and recognitions.

What started out as a job criss-crossing the United States and Canada selling pipeline and cable markers in the early 1980's turned into a passion for helping utilities prevent damage to their buried facilities ensuring contractors can dig safely. Damage prevention got into Scott's blood – attending conferences with 811 Centers and utilities all working to make excavation safe. This led Scott to write articles for many industry publications and become very involved with the Common Ground Alliance, including serving as co-chair of the Education Committee for four years and receiving the CGA's Jim Barron Award in 2007. The Jim Barron Award recognizes an individual or organization that has shown exceptional, sustained dedication to improving underground utility damage prevention.

Scott would go on to start his own damage prevention products company as well as a conference, training, and media company focused solely on damage prevention. His company created the world's largest damage prevention conference and co-founded the first damage prevention conference in Australia. He would go on to hold 14 patents, including 10 for damage prevention related marking products, and deliver speaking presentations on five continents. This industry is full of people passionate about making a difference and saving lives including 811 Center professionals, engineers, public awareness professionals, locators, and many more dedicated professionals. As Scott's career has shown, even the people selling products and services can have a significant impact on public safety.

Only 38 years behind Scott, Benjamin's story is only just getting started (a spry 30 years old at the start of this book-writing project). His entry point to damage prevention is worth noting here as well. Studying first economics and pursuing further study in public policy and then in law, his aim was always to work in a think tank. But from this particular perch, Benjamin was able to study and survey public policy and the various laws and regulations that form, guide, and structure the damage prevention industry. In keeping with the goal that more people can and should enter the damage prevention field, Benjamin's path highlights a notable need: policy reform. From his vantage point outside the industry, Benjamin is the first to admit he isn't a stakeholder and recognizes the need to listen to the voices of the veterans. But as a third-party researcher and educator, he has also helped identify blind spots and put forward resources for industry participants to consider and build off of.

The future of damage prevention absolutely requires new technology, new best practices, and improved communication and collaboration between and among stakeholders. It also needs people thinking about and interfacing with the public policy process to streamline the rules, tighten them in places, and remove them in others. It will require lawyers and insurers, investigators and educators, and even more. The direct and supporting roles in and around the industry are all important. So whether someone reading this book intends to go into excavating, locating, utility work, or even aspires to public office, they all have an entry point to work in damage prevention and to bring about the next great leap in public safety, infrastructure resilience, and economic prosperity.

The industry has come a long way and improved both public safety and excavator safety. Now it is up to the next generation of damage prevention professionals to continue those steps further forward.

While this book doesn't address every nuance in the industry, it is a glimpse behind the scenes of everyday life and an insight into the bombs that never go off. Spotlighting this industry for its own merit, not as a microscope after an incident is critical, because the men and women doing this work do so tirelessly and often thanklessly. A short overview of what they do, why it matters, and how it is evolving only scratches the surface. Indeed, the deeper you go, the more interconnected and complex the dynamics are. This project is merely an introduction for most and potentially a helpful reframing for others. The rest is yet to be written *by you.*

Afterword by Bob Kipp

If you have read this far, you now understand something most people never fully grasp, how much we all depend on what buried beneath us.

Every glass of clean water.

Every warm home in winter.

Every hospital, airport, data center, school, and neighborhood.

All of it depends on pipelines, power lines, fiber networks, and facilities that quietly make modern life possible. When they work, no one notices. When they fail, everyone does, and sometimes people are injured and some die. We can help prevent this from happening.

This book has chronicled the people, systems, and innovations that hold back disaster every single day. It has shown how One Call systems, locators, engineers, excavators, regulators, and educators form a chain of responsibility. A chain that protects lives, communities, and economies.

Having worked for Bell Canada for many years, I had plenty of experience dealing with cut cables and the repercussions. When I became the Executive Director of the Common Ground Alliance, my perspective became much broader. The small CGA staff and all our committed members worked hard to improve safety. When we launched 811 in 2007, everything changed. As I recall, in a few years, damages were down by nearly 40%.

But the story is not finished. The next chapter belongs to the next generation.

To young people searching for a career that has a positive impact, this industry offers something rare: the ability to protect our vital infrastructure and even save lives. Damage prevention is not abstract. It is not theoretical. It is hands-on, technology-driven, community-centered, and mission-critical. It blends mapping, data science, engineering, public safety, field operations, communications, and leadership. It rewards problem-solvers. It needs innovators. It depends on the people who care.

If you are early in your career or guiding someone who is, the damage prevention industry is rewarding and full of opportunity. It offers stability, advancement, and the chance to make a tangible difference every single day. You can build systems that protect cities. You can develop technologies that reduce risk. You can train crews that prevent damages, outages, and injuries. You can quite literally help hold back disaster.

And to everyone, whether you work in the industry or simply benefit from it, do not take it for granted. Before any shovel breaks soil, before any post is set, before any trench is opened, there is one simple action that protects everything described in this book: have all the buried facilities located.

Safe digging is not bureaucracy. It is not delay. It is responsibility. It is how we prevent tragedies before they happen. It is how we protect workers and families. It is how we preserve the invisible network that sustains modern life.

The systems described in these pages were built by people who refused to accept preventable damage as inevitable. They collaborated. They improved. They educated. They measured. They adapted.

Now it is your turn.

May this book serve not just as a history, but as a call to lead, to innovate, to recruit, to educate, and to dig with care. After having the privilege to serve as the CGA, I can guarantee that if you join this industry, you will never regret it, and you will make a difference.

Robert "Bob" Kipp served as President of the Common Ground Alliance (CGA) from the organization's inception in 2001 until 2017. As a Canadian telecommunications executive and long-time industry stakeholder in underground infrastructure protection, he helped advance collaboration, cross-border cooperation, and best practices for damage prevention across the United States and Canada.

From Scott

In my 40+ years in the damage prevention industry, the one constant has been the passion of the professionals in the industry. The biggest changes in the industry have been improved communication among stakeholders, better-designed targeted education, and technology improvement within the 811 Centers. I am very excited to see how the combination of rapid improvements in technology and the new ideas and fresh perspectives of the young people coming into the industry combine to reduce damages radically. I believe the development and implementation of an industry-wide damage prevention certification program will be the most significant improvement to the industry in the next 10 years. A certification program will draw more top-notch people to all segments of the industry and lead to a steady decrease in damages.

From Benjamin

It is my hope that within five years, major portions of this book will be obsolete due to major reforms, and in 15 years, it is going to be a relic of history describing a bygone era. I hope the next generation takes up the mantle of damage prevention and infuses new energy, technology, and perspective into the system – while learning from and complementing the hard-won wisdom of the veterans currently holding this industry up. And finally, I hope our smarter and interconnected infrastructure helps protect itself through prudent investment in concepts like distributed fiber optic sensing and other applications providing real-time intelligence and situational awareness. When every excavator goes home to his family every night, and every family at home has reliable public utility service – without disruption – damage prevention will be fully realized.

Glossary of Terms

Abandoned Facility/Line A buried utility that is no longer in service but still present underground; may cause hazards or confusion during excavation if not properly identified

Brand Ambassador An individual who promotes the safe excavation and 811 message to the public, often through industry leadership, speaking, or creative outreach campaigns

Cross Bore An intersection of a new utility line with an existing underground utility, often undetected, which can create significant safety hazards

Damage Prevention The system of practices, regulations, and education aimed at preventing excavation-related damage to underground infrastructure

Damage Prevention Program A set of required or voluntary practices (sometimes mandated by regulation) that utilities must implement to minimize risks during excavation

Excavation Safety Practices that protect workers and the public from hazards during digging and construction activities

Locator A trained professional who identifies and marks the location of underground facilities using specialized equipment

One-Call Ticket The official notice created when an excavator contacts 811; it notifies utilities of planned excavation and initiates the locate process

One-Call Center This is used both with and without a hyphen by different industry stakeholders. These may also be referred to as a "Facility Notification Center" or "811 Center." These are regional or state-run nonprofit organizations that receive excavation notices and notify facility operators to mark underground lines

Potholing/Daylighting Safe digging techniques (such as vacuum excavation or hand-digging) used to physically expose a buried utility and confirm its location

Positive Response An official communication from facility operators or locators to the excavator confirming that utilities have been marked, or that no facilities are present at the site

Right-of-Way A legal corridor or easement allowing infrastructure such as pipelines, cables, or roads to be installed, accessed, and maintained across a defined area of land

Stakeholder Groups Industry sectors with a role in damage prevention, including excavators, locators, utilities, regulators, insurers, telecom, and others

Subsurface Utility Engineering (SUE) A professional engineering practice that investigates and maps underground utilities to reduce conflicts during construction

Tolerance Zone The buffer area on either side of a marked underground facility within which special care must be taken when excavating; the width varies by state law

Utility Coordinating Committee (UCC) Local or regional groups organized to improve communication and coordination between utilities, excavators, and public agencies

White Lining The practice of marking a proposed excavation area with white paint or flags before an official locate is performed

About the Authors

Scott and Benjamin were first introduced virtually by Mike Sullivan on a video call in April 2023. The first in-person meeting occurred in Banff in February 2024 at the Utility Safety Partners 40th Anniversary conference, where Benjamin presented industry research and survey results. Many phone calls, emails, video calls, and conference catch ups ensued.

Scott is a longtime writer, publisher, inventor, entrepreneur, and speaker who has been involved in the damage prevention industry since the early 1980s.

Benjamin is the executive director of the Alliance for Innovation and Infrastructure, the nation's only think tank dedicated to infrastructure.

Index

For Product Safety Concerns and Information please contact our EU representative GPSR@taylorandfrancis.com
Taylor & Francis Verlag GmbH, Kaufingerstraße 24, 80331 München, Germany

www.ingramcontent.com/pod-product-compliance
Lightning Source LLC
LaVergne TN
LVHW081316110826
845149LV00006B/1518

* 9 7 8 1 0 4 1 3 5 1 2 1 4 *